# Cerebral Ventricles

Hayder R. Salih · Samer S. Hoz
Ali A. Dolachee · Mohammed A. Alrawi
Zaid Aljuboori · Mayur Sharma
Mustafa Ismail · Norberto Andaluz

*Editors*

# Cerebral Ventricles

In Multiple-Choice Questions

*Editors*
Hayder R. Salih
Department of Neurosurgery
Neurosurgery Teaching Hospital
Baghdad, Iraq

Ali A. Dolachee
Department of Surgery
University of Baghdad
Baghdad, Iraq

Zaid Aljuboori
Department of Neurosurgery
Geisinger Commonwealth School of
Medicine
Scranton, PA, USA

Mustafa Ismail
Department of Neurosurgery
Neurosurgery Teaching Hospital
Baghdad, Iraq

Samer S. Hoz
Department of Neurosurgery
University of Cincinnati
Cincinnati, OH, USA

Mohammed A. Alrawi
Department of Neurosurgery
Neurosurgery Teaching Hospital
Baghdad, Iraq

Mayur Sharma
Department of Neurosurgery
University of Minnesota
Minnesota, MN, USA

Norberto Andaluz
Department of Neurosurgery
University of Cincinnati
Cincinnati, OH, USA

ISBN 978-3-031-42341-3        ISBN 978-3-031-42342-0   (eBook)
https://doi.org/10.1007/978-3-031-42342-0



This Springer imprint is published by the registered company Springer Nature Switzerland AG
The registered company address is: Gewerbestrasse 11, 6330 Cham, Switzerland

Paper in this product is recyclable.

To the memory of my dear father who was a role model for me in searching for knowledge and imparting it to his students. To our teachers and mentors who enlightened us on the path of neurosurgery.
**Hayder R. Salih**

To my lovely family: Sawsan, Saad, Samher, Arwa, Farah, Ward, and Sanaa.
To Prof. A. Hadi Al Khalili, Paolo Palmisciano, and the best supporter Kathleen Smith.
**Samer S. Hoz**

To my lovely family which supports me in this long way, neurosurgery.
**Ali A. Dolachee**

To my patients, mentors and to my family.
**Mohammed A. Alrawi**

I would like to dedicate this book to my mentors for their support and guidance.
**Zaid Aljuboori**

To my mentors, parents (mother and late father), wife (Smita), daughters (Riya and Shreya), friends, and my patients who have given me the opportunity to learn.
**Mayur Sharma**

For my father's memory. May God have mercy upon your soul.
**Mustafa Ismail**

To my family, my patients, and my trainees. You make it all worthwhile.
**Norberto Andaluz**

# Foreword 1

When the authors kindly invited me to write a few words as a foreword to their book *Cerebral Ventricles in Multiple Choice Questions* I asked them for time to read it and evaluate their work. After a few days, I found on a gentle rainy Saturday the time to review the text of this beautiful book. I quickly got caught up in the new Multiple-Choice format, where contrary to what we know, the correct answer was to identify the false answer. I discovered in this format that the reader has the possibility to remember in the four options that are presented as correct the main characteristics of an anatomical region, pathology, or a surgical approach related to the cerebral ventricles. So, whoever reads this book is acquiring, remembering, or reaffirming knowledge as they read the questions. You do not have to find a correct answer among four false ones, but on the contrary, you will be reading four truths about a topic, and you must discover the false one. In the clarification that is made of the correct answer, which paradoxically turns out to be the only false premise, the reader also learns by just reading the explanation at the bottom of each question without having to turn the pages. I think that the Multiple-Choice format in a certain way "inverted" where the false is the correct one is the most novel thing about this book, and this is the reason that makes it easy to read and a tool to refresh everything that a neurosurgeon should know about the cerebral ventricles. I found it fascinating. The other evaluation that must be made is regarding the contents, the way and the order in which they are presented, and the arrangement in which they are exposed in the book. I want to congratulate the authors because they abound in detail with great precision in each of the contents with an excellent power of synthesis and hierarchy of the most important contents for the Neurosurgeon, for the anatomist or for anyone who is interested in knowing in detail the ventricular encephalic system. I congratulate the authors and know that many Neurosurgeons, fellows, students, and general practitioners will enjoy reading this book.

**Roberto R. Herrera**
Buenos Aires, Argentina

# Foreword 2

The editor of this book represents a vibrant group of neurosurgeons, who are exemplary in their specialized medical service to their community under strenuous circumstances. They strive to share their knowledge, through the dissemination of high-quality well-structured review books; with endless numbers of neurosurgeons at various levels of their training and career. Their approach is simple yet effective. Get on a focused subject, in this case, the cerebral ventricles; dissect it very well, and put it together in an elegant way formed as questions and answers reflecting their deep understanding of the matter at hand. Very few books truly achieve this goal. This book is exemplary in giving the readership a solid base to stand on when approaching the biology and pathophysiology of the cerebral ventricles. This will guide many neurosurgeons through their reviews, making it through their rough transitions (exams) and beyond (career). This is a great effort for which the authors must be commended.

**Hosam Al-Jehani**
Imam Abdulrahman Alfaisal University
Dammam, Saudi Arabia

Montreal Neurological Institute and Hospital
McGill University
Montreal, Canada

Adjunct Assistant Professor, Department of Neurosurgery
Weill Cornell University
Houston Methodist, USA

# Preface

Dear reader

It is with profound privilege that we present this meticulously crafted volume on one of the most complex, challenging, and fascinating structures of our brains, the cerebral ventricles. The cerebral ventricles, as an integral component of the human central nervous system, possess an intricate anatomy, complex physiology, and relationships with the surrounding critical neurovascular structures of the brain.

The lesions involving the cerebral ventricles cause significant distortion of the already complex anatomy of the ventricles. Therefore, in navigating the labyrinthine complexities of cerebral ventricular lesions and their surgical management, one must tread with precision and a profound understanding of the anatomy is critical. This is our humble attempt to elucidate the intricate complexities of these most enigmatic and crucial facets of the human brain and neurosurgical practice related to the surgical management of lesions involving this complex structure.

In this endeavor, we have employed a series of concise multiple choice questions, designed to instill robust, concrete knowledge in students, residents, fellows, and junior neurosurgeons. Whether you are preparing for rigorous board examinations or stepping into clinical practice, we envision this book to be an invaluable tool along your journey.

We sincerely hope this book enriches your understanding and sparks your curiosity, and we wish you a deeply rewarding and insightful experience as you delve into this book.


**Hayder R. Salih**
Baghdad, Iraq
Cincinnati, OH, USA
Baghdad, Iraq
Baghdad, Iraq
Scranton, PA, USA
Minnesota, MN, USA
Baghdad, Iraq
Cincinnati, OH, USA

Samer S. Hoz
Ali A. Dolachee
Mohammed A. Alrawi
Zaid Aljuboori
Mayur Sharma
Mustafa Ismail
Norberto Andaluz

# Key Features

- Cerebral Ventricles in Multiple Choice Questions is the FIRST review book to use the multiple choice question (MCQ) format to specifically address the topic of cerebral ventricles.
- The mission of this book is to help readers revise the core concepts and maintain knowledge of the anatomy, pathology, and surgery of cerebral ventricles.
- This study companion is structured in five sections, for a total of 18 chapters, including 553 MCQs in a convenient format that is suitable for self-study.
- Answers and explanations appear immediately below the questions to enhance readability.
- This book is an adjunct to existing texts and does not intend to be the primary source of information; it rather aims to help readers identify their relevant strengths and weaknesses in the area.
- Cerebral Ventricles in Multiple Choice Questions is based on the most up-to-date best practice evidence, with a style that mirrors the format adopted by the majority of local, regional, and international board examinations.
- The student of neurosurgery, the resident, the fellow, the younger neurosurgeon preparing for exams or practice, and even the later stage neurosurgeon are the target audience of this book.
- This topic-focused MCQ book follows the concept and style of the internationally recognized book *Vascular Neurosurgery in Multiple Choice Questions*, authored by Dr. Samer S. Hoz and published by Springer 2017. This book receives an esteemed scientific review in Neurosurgery Journal and is a recommended book for the American Board of Neurological Surgery (ABNS). On the same pathway, our second book in the series *Neurotrauma in Multiple Choice Questions* was published by Springer 2021 and got another scientific book review in Neurosurgery Journal as well.

# Contents


**1 Lateral Ventricle: Anatomy** ........................................................... 1

Hayder R. Salih, Mustafa Ismail, Saleh A. Saleh,
Younus M. Al-Khazaali, Sura H. Talib, and Samer S. Hoz

**2 Lateral Ventricle: Pathology: Tumors** ........................................... 9

Abdullah H. Al Ramadan, Sarah A. Basindwah, Sadeem A. Albulaihed,
Abdulaziz Y. Alahmed, Abdulaziz S. Alayyaf, Wafa M. Almuallim,
and Mustafa Ismail

**3 Lateral Ventricle: Pathology: Non-Tumors** ................................... 23

Maliya Delawan, Younus M. Al-Khazaali, Soubhagya R. Tripathy,
Aanab O. Abdulameer, Ali A. Dolachee, and Mohammed A. Alrawi

**4 Lateral Ventricle: Surgery** ............................................................. 35

Viraat Harsh, Manoj Kumar, Anil Kumar, Margarida S. Conceicao,
Osama M. Al-Awadi, Mustafa Ismail, and Samer S. Hoz

**5 Third Ventricle: Anatomy** .............................................................. 47

Hayder R. Salih, Margarida S. Conceicao, Mostafa H. Algabri,
Mahmood F. Alzaidy, Ibrahim A. Izzet, Zaid Aljuboori, and Samer S. Hoz

**6 Third Ventricle: Pathology—Tumors** ............................................. 57

Baha'eddin A. Muhsen, Hazem Madi, Mustafa Ismail,
Ignatius N. Esene, and Norberto Andaluz

**7 Third Ventricle: Pathology—Non-Tumors** ..................................... 67

Hernando Alvis-Miranda, Teeba A. Al-Ageely, Aqeel K. Ibrahim,
Salam N. Hussein, Mostafa H. Algabri, Gulshan T. Muhammed,
and Mustafa Ismail

**8 Third Ventricle: Surgery** ................................................................ 89

Waeel O. Hamouda, Maliya Delawan, Aktham O. Al-Khafaji,
and Mayur Sharma

**9 Fourth Ventricle: Anatomy** ........................................................................ 117

Hayder R. Salih, Maria L. Laffitte, Naba G. Husseini,
Saleh A. Saleh, Mohammed S. Altaee, and Samer S. Hoz

**10 Fourth Ventricle: Pathology—Tumors** ...................................................... 125

Ali A. Dolachee, Abdullah K. Al-Qaraghuli, Sajjad G. Al-Badri,
Alkawthar M. Abdulsada, Hagar A. Algburi, Zinah A. Al-araji,
and Mustafa Ismail

**11 Fourth Ventricle: Pathology—Non-Tumors** ............................................. 141

Rupesh Raut, Mohamed M. Arnaout, Ali M. Neamah,
Saja A. Albanaa, Noor M. Akar, and Mustafa Ismail

**12 Fourth Ventricle: Surgery** ........................................................................... 149

Waeel O. Hamouda, Yara Alfawares, Vishan P. Ramanathan,
and Mustafa Ismail

**13 CSF and Hydrocephalus** ............................................................................. 169

Eleni Tsianaka, Sara A. Mohammad, Mustafa Ismail,
and Awfa Aktham

**14 CSF Diversion** ............................................................................................... 175

Laith T. Al-Ameri, Noor F. Hassan, Ali Q. Alforati,
and Mustafa A. Shamkhi

**15 Cerebral Ventricle: Congenital Lesions** .................................................... 185

Serge E. Mba, Fatimah O. Ahmed, Ahmed I. Abdulwahhab, Rania H.
Al-Taie, Zahraa A. Alsubaihawi, and Mohammed Al-Dhahir

**16 Cerebral Ventricle: Infection** ..................................................................... 195

Muthanna N. Abdulqader, Ruqaya A. Kassim, Hashim T. Hashim,
Mustafa E. Almurayati, Nabeel H. Al-Fatlawi,
and Mustafa Ismail

**17 Cerebral Ventricle: Vascular Anatomy** ...................................................... 205

Hayder R. Salih, Hatem Sadik, Sama Albairmani, Mustafa Ismail,
and Samer S. Hoz

**18    Cerebral Ventricle: Vascular Lesions and Hemorrhage**...................... 213
*Osman Elamin, Ahmed Muthana, Hussein M. Hasan,
Wamedh E. Matti, Hussein J. Kadhum, and Samer S. Hoz*

**References** ................................................................................. 222

# Contributors

**Aanab O. Abdulameer** College of Medicine, University of Baghdad, Baghdad, Iraq

**Muthanna N. Abdulqader** Department of Neurosurgery, Neurosurgery Teaching Hospital, Baghdad, Iraq

**Alkawthar M. Abdulsada** Azerbaijan Medical University, Baku, Azerbaijan

**Ahmed I. Abdulwahhab** Al-Kindy College of Medicine, University of Baghdad, Baghdad, Iraq

**Fatimah O. Ahmed** College of Medicine, University of Baghdad, Baghdad, Iraq

**Noor M. Akar** College of Medicine, Al-Nahrain University, Baghdad, Iraq

**Awfa Aktham** Department of Neurosurgery, Tokyo General Hospital, Nakano, Japan

**Teeba A. Al-Ageely** College of Medicine, University of Baghdad, Baghdad, Iraq

**Abdulaziz Y. Alahmed** Department of Neurosurgery, College of Medicine, King Faisal University, Alahsa, Saudi Arabia

**Laith T. Al-Ameri** Al-Kindy College of Medicine, University of Baghdad, Baghdad, Iraq

**Zinah A. Al-araji** College of Medicine, Alnahrain University, Baghdad, Iraq

**Osama M. Al-Awadi** Department of Neurosurgery, Neurosurgery Teaching Hospital, Baghdad, Iraq

**Abdulaziz S. Alayyaf** College of Medicine, Prince Sattam Bin Abdulaziz University, Riyadh, Saudi Arabia

**Sajjad G. Al-Badri** College of Medicine, University of Baghdad, Baghdad, Iraq

**Sama Albairmani**   College of Medicine, University of Al-Iraqia, Baghdad, Iraq

**Saja A. Albanaa**   College of Medicine, University of Baghdad, Baghdad, Iraq

**Sadeem A. Albulaihed**   College of Medicine, Alfaisal University, Riyadh, Saudi Arabia

**Mohammed Al-Dhahir**   Department of Neurosurgery, Yemeni German Hospital, Sana'a, Yemen

**Nabeel H. Al-Fatlawi**   College of Medicine, University of Baghdad, Baghdad, Iraq

**Yara Alfawares**   University of Cincinnati College of Medicine, Cincinnati, OH, USA

**Ali Q. Alforati**   College of Medicine, University of Baghdad, Baghdad, Iraq

**Mostafa H. Algabri**   College of Medicine, University of Baghdad, Baghdad, Iraq

**Hagar A. Algburi**   College of Medicine, University of Baghdad, Baghdad, Iraq

**Zaid Aljuboori**   Department of Neurosurgery, Geisinger Commonwealth School of Medicine, Scranton, PA, USA

**Aktham O. Al-Khafaji**   Department of Surgery, College of Medicine, University of Baghdad, Baghdad, Iraq

**Younus M. Al-Khazaali**   College of Medicine, Al-Nahrain University, Baghdad, Iraq

**Wafa M. Almuallim**   College of Medicine, Madinah, Saudi Arabia

**Mustafa E. Almurayati**   College of Medicine, University of Baghdad, Baghdad, Iraq

**Abdullah K. Al-Qaraghuli**   MedStar Health Research Institute, MedStar Washington Hospital Center, Washington, DC, USA

**Mohammed A. Alrawi**   Department of Neurosurgery, Neurosurgery Teaching Hospital, Baghdad, Iraq

**Zahraa A. Alsubaihawi**  College of Medicine, University of Baghdad, Baghdad, Iraq

**Mohammed S. Altaee**  Neurosurgery Department, Azadi Teaching Hospital, Kirkuk, Iraq

**Rania H. Al-Taie**  College of Medicine, University of Mustansiriyah, Baghdad, Iraq

**Hernando Alvis-Miranda**  Colombian Foundation Center for Epilepsy and Neurological Diseases—FIRE, Bolívar, Colombia

**Mahmood F. Alzaidy**  College of Medicine, University of Baghdad, Baghdad, Iraq

**Norberto Andaluz**  Department of Neurosurgery, University of Cincinnati College of Medicine, Cincinnati, OH, USA

**Mohamed M. Arnaout**  Department of Neurosurgery, Zagazig University, Zagazig, Egypt

**Sarah A. Basindwah**  Department of Neurosurgery, King Khalid University Hospital at King Saud University, Riyadh, Saudi Arabia

**Margarida S. Conceicao**  CHTV—Hospital São Teotónio, University of Porto, Porto, Portugal

**Maliya Delawan**  College of Medicine, Gulf Medical University, Ajman, United Arab Emirates

**Ali A. Dolachee**  Department of Surgery, Al-Kindy College of Medicine, University of Baghdad, Baghdad, Iraq

**Osman Elamin**  Department of Neurosurgery, Jordan Hospital and Medical Center, Amman, Jordan

**Ignatius N. Esene**  Neurosurgery Division, Faculty of Health Sciences, University of Bamenda, Bambili, Cameroon

**Waeel O. Hamouda**  Neurological and Spinal Surgery Department, Cairo University School of Medicine and Teaching Hospitals, Cairo, Egypt

Cairo University, Faculty of Medicine, Cairo, Egypt

**Viraat Harsh**  Department of Neurosurgery, Rajendra Institute of Medical Sciences, Ranchi, Jharkhand, India

**Hussein M. Hasan**  College of Medicine, University of Baghdad, Baghdad, Iraq

**Hashim T. Hashim**  College of Medicine, University of Baghdad, Baghdad, Iraq

**Noor F. Hassan**  Department of Neurosurgery, Neurosurgery Teaching Hospital, Baghdad, Iraq

**Samer S. Hoz**  Department of Neurosurgery, University of Cincinnati, Cincinnati, OH, USA

**Naba G. Husseini**  College of Medicine, University of Mosul, Mosul, Iraq

**Salam N. Hussein**  Department of Neurosurgery, Neurosurgery Teaching Hospital, Baghdad, Iraq

**Aqeel K. Ibrahim**  Department of Neurosurgery, Neurosurgery Teaching Hospital, Baghdad, Iraq

**Mustafa Ismail**  Department of Neurosurgery, Neurosurgery Teaching Hospital, Baghdad, Iraq

**Ibrahim A. Izzet**  College of Medicine, University of Baghdad, Baghdad, Iraq

**Hussein J. Kadhum**  Department of Neurosurgery, Neurosurgery Teaching Hospital, Baghdad, Iraq

**Ruqaya A. Kassim**  College of Medicine, University of Baghdad, Baghdad, Iraq

**Anil Kumar**  Department of Neurosurgery, Rajendra Institute of Medical Sciences, Ranchi, Jharkhand, India

**Manoj Kumar**   Department of Neurosurgery, Taha Motors Hospital, Jamshedpur, Jharkhand, India

**Maria L. Laffitte**   Salford Royal Hospital NHS Foundation Trust, Manchester, UK

**Hazem Madi**   European Gaza Hospital, Gaza, Palestine

**Wamedh E. Matti**   Department of Neurosurgery, Neurosurgery Teaching Hospital, Baghdad, Iraq

**Serge E. Mba**   Spine and Brain Surgery Centre, Harare, Zimbabwe

**Sara A. Mohammad**   College of Medicine, University of Baghdad, Baghdad, Iraq

**Gulshan T. Muhammed**   Department of Neurosurgery, Neurosurgery Teaching Hospital, Baghdad, Iraq

**Baha'eddin A. Muhsen**   Department of Neurosurgery, Islamic Hospital, Amman, Jordan

**Ahmed Muthana**   College of Medicine, University of Baghdad, Baghdad, Iraq

**Ali M. Neamah**   College of Medicine, University of Baghdad, Baghdad, Iraq

**Abdullah H. Al Ramadan**   Pediatric Neurosurgery Department, Maternity and Children Hospital, First Health Cluster, Eastern Province, Saudi Arabia

**Vishan P. Ramanathan**   University of Cincinnati College of Medicine, Cincinnati, OH, USA

**Rupesh Raut**   Department of Neurosurgery, Patan Academy of Health Sciences, Patan Hospital, Kathmandu, Nepal

**Hatem Sadik**   Nuffield Department of Surgical Sciences, University of Oxford, Oxford, UK

**Saleh A. Saleh**   College of Medicine, University of Baghdad, Baghdad, Iraq

**Hayder R. Salih**   Department of Neurosurgery, Neurosurgery Teaching Hospital, Baghdad, Iraq

**Mustafa A. Shamkhi**   Department of Neurosurgery, Neurosurgery Teaching Hospital, Baghdad, Iraq

**Mayur Sharma**   Department of Neurosurgery, University of Minnesota, Minneapolis, MN, USA

**Sura H. Talib**   College of Medicine, Al-Mustansiriyah University, Baghdad, Iraq

**Soubhagya R. Tripathy**   Department of Neurosurgery, Hitech Medical College, Bhubaneswar, India

**Eleni Tsianaka**   Neurosurgery Department, Kuwait Hospital, Sabah Al Salem, Kuwait

# Abbreviations

| | |
|---|---|
| **AChA** | Anterior choroidal arteries |
| **AICA** | Anterior inferior cerebellar artery |
| **BBB** | Blood–brain barrier |
| **CNS** | Central nervous system |
| **CSF** | Cerebrospinal fluid |
| **CT** | Computed tomography |
| **DWI** | Diffusion-weighted imaging |
| **EEG** | Electroencephalography |
| **ETV** | Endoscopic third ventriculostomy |
| **EVD** | External ventricular drain |
| **FLAIR** | Fluid attenuation inversion recovery |
| **GFAP** | Glial fibrillary acidic protein |
| **HH** | Hypothalamic hamartomas |
| **ICA** | Internal carotid artery |
| **ICH** | Intracerebral hemorrhage |
| **ICP** | Intracranial pressure |
| **INR** | International normalized ratio |
| **IVH** | Intraventricular hemorrhage |
| **IVH-n** | Intraventricular hemorrhage-in newborn |
| **MB** | Medulloblastoma |
| **MBEN** | Medulloblastoma with extensive nodularity |
| **MRSA** | Methicillin-resistant *Staphylococcus aureus* |
| **NPH** | Normal pressure hydrocephalus |
| **PICA** | Posterior inferior cerebellar artery |
| **SAH** | Subarachnoid hemorrhage |
| **SEGA** | Subependymal giant cell astrocytoma |
| **SHH** | Sonic hedgehog |
| **SSEP** | Somatosensory-evoked potentials |
| **SSS** | Superior sagittal sinus |

| | |
|---|---|
| **T1WI** | T1-weighted image |
| **T2WI** | T2-weighted image |
| **VHL** | Von Hippel-Lindau (disease) |
| **VP** | Ventriculoperitoneal |
| **WNT** | WNT/Wingless |

# Lateral Ventricle: Anatomy

*Hayder R. Salih, Mustafa Ismail, Saleh A. Saleh, Younus M. Al-Khazaali, Sura H. Talib, and Samer S. Hoz*

**1.** Lateral ventricle compartments, the FALSE answer is:
   A. Occipital horn.
   B. Frontal horn.
   C. Antrum.
   D. Temporal horn.
   E. Body.

**Answer C**
   - The atrium is a part of the lateral ventricle.
   - Each lateral ventricle has five parts: the frontal, temporal, and occipital horns, the body, and the atrium. Each of these five parts has medial and lateral walls, a roof, and a floor. In addition, the frontal and temporal horns and the atrium have anterior walls.

**2.** Thalamus, the FALSE answer is:
   A. The body of the lateral ventricle is above the thalamus.
   B. The atrium and the occipital horn are posterior to the thalamus.
   C. The temporal horn is infero-lateral to the thalamus.
   D. The medial part of the thalamic pulvinar forms the anterior wall of the atrium.
   E. The inferior surface of the thalamus forms the floor of the body.

**Answer E**
   - The superior surface of the thalamus forms the floor of the body.
   - The inferior surface of the thalamus forms the medial edge of the roof of the temporal horn.

**3.** Caudate nucleus, the FALSE answer is:
   A. Caudate nucleus is a C-shaped mass that wraps around the thalamus.
   B. The caudate head bulges into the lateral wall of the frontal horn and body.
   C. The caudate body forms part of the lateral wall of the atrium.
   D. The caudate tail extends from the atrium into the roof of frontal horn.
   E. The caudate tail is continuous with amygdala at roof of occipital horn.

**Answer D**
   - The tail extends from the atrium into the roof of the temporal horn.

**? 4. Caudate nucleus and thalamus, the FALSE answer is:**
A. In the body of lateral ventricle, the caudate is supero-lateral to the thalamus.
B. In the atrium, the caudate is medial to the thalamus.
C. In the temporal horn, the caudate is infero-lateral to the thalamus.
D. The stria terminalis runs between the caudate and the thalamus.
E. The stria terminalis runs parallel and deep to the thalamo-striate vein.

**✓ Answer B**
– In the atrium, the caudate is postero-lateral to the thalamus.

**? 5. Fornix, the FALSE answer is:**
A. It is a C-shaped structure that wraps around the thalamus.
B. It consists mainly of hippocampo-mamillary tract fibers.
C. The fibers originate from hippocampus, subiculum, and dentate gyrus of temporal lobe.
D. The fimbria of fornix arises from the floor of temporal horn of ventricular surface of hippocampus.
E. The fimbria passes anteriorly to become the column of the fornix.

**✓ Answer E**
– The fimbria passes posteriorly to become the crus of the fornix.

**? 6. Fornix, the FALSE answer is:**
A. The crus of fornix wrap around posterior surface of the pulvinar of thalamus.
B. The body of fornix runs along supero-medial border of thalami in medial wall of body of lateral ventricle.
C. The body of fornix separates the roof of third ventricle from the floor of lateral ventricle.
D. The body of fornix separates into two columns at the posterior margin of the thalamus.
E. Below splenium, the two crura of fornix interconnect to form the hippocampal commissure.

**✓ Answer D**
– The body of the fornix separates into two columns at the anterior margin of the thalamus that arch along the anterior and superior margin of the foramen of Monro.

**7. Fornix, the FALSE answer is:**
A. The body of fornix is in the lower part of the medial wall of the body of lateral ventricle.
B. The crus of fornix is in the medial part of the anterior wall of the atrium.
C. The fimbria of fornix is in the medial part of the floor of the occipital horn.
D. The body of fornix crosses the midway of superior surface of thalamus.
E. The crus of fornix crosses the midway of the pulvinar.

**Answer C**
- The fimbria of the fornix is in the medial part of the floor of the temporal horn.

**8. Fornix, the FALSE answer is:**
A. Part of thalamus lateral to the body of the fornix forms the roof of the body of lateral ventricle.
B. Part of thalamus medial to the body of the fornix forms part of the lateral wall of velum interpositum and third ventricle.
C. Part of pulvinar lateral to crus of fornix forms part of the anterior wall of atrium.
D. Part of pulvinar medial to fornix forms part of anterior wall of quadrigeminal cistern.
E. Part of thalamus medial to the fimbria forms the roof of the ambient cistern.

**Answer A**
- The part of the thalamus lateral to the body of the fornix forms the floor of the body of the lateral ventricle.

**9. Corpus callosum, the FALSE answer is:**
A. Rostrum forms the floor of the frontal horn.
B. The forceps minor of genu forms anterior wall of the frontal horn.
C. Genu and body form the roof of both frontal horn and the body of lateral ventricle.
D. The forceps major of splenium forms part of medial wall of atrium and occipital horn.
E. Tapetum forms the floor of the frontal horns.

✅ **Answer E**
- The tapetum forms the roof and lateral wall of the atrium, temporal and occipital horns.

❓ **10. Septum pellucidum, the FALSE answer is:**
   A. Composed of single lamina.
   B. Separates the frontal horns and bodies of lateral ventricles.
   C. Is tallest anteriorly and shortest posteriorly.
   D. It disappears near the junction of the body and crura of the fornix.
   E. The antero-posterior length ranges from 28 to 50 mm.

✅ **Answer A**
- Composed of paired laminae.
- There may be a cavity in the midline between the laminae (cavum septum pellucidum).

❓ **11. Septum pellucidum attachments, the FALSE answer is:**
   A. To the rostrum of the corpus callosum below.
   B. To the genu of the corpus callosum anteriorly.
   C. To the splenium of the corpus callosum anteriorly.
   D. To the body of the corpus callosum above.
   E. To the body of the fornix posteriorly.

✅ **Answer C**
- To the splenium of the corpus callosum caudally (posteriorly).

❓ **12. Internal capsule, the FALSE answer is:**
   A. The anterior limb is located between caudate and lentiform nuclei.
   B. The anterior limb is separated from the frontal horn by the head of caudate nucleus.
   C. The posterior limb is located between the claustrum and lentiform nucleus.
   D. The posterior limb is separated from the body by the thalamus and the caudate.
   E. Genu touches the wall of lateral ventricle lateral to the foramen of Monro.

✅ **Answer C**
- The posterior limb of the internal capsule is located between the thalamus and lentiform nucleus.

**13. Lateral ventricular walls, the frontal horn, the FALSE answer is:**
   A. The medial wall is formed by the septum pellucidum.
   B. The anterior wall and roof are formed by the genu of corpus callosum.
   C. The lateral wall is formed by the head of the caudate nucleus.
   D. The floor is formed by the rostrum of the corpus callosum.
   E. The columns of the fornix form the superior part of lateral wall.

✓ **Answer E**
   – The columns of the fornix form the postero-inferior part of the medial wall.

**14. Lateral ventricular walls, the body, the FALSE answer is:**
   A. The roof is formed by the body of the corpus callosum.
   B. The medial wall is formed by the septum pellucidum above.
   C. The medial wall is formed by the fimbria of fornix.
   D. The lateral wall is formed by the body of the caudate nucleus.
   E. The floor is formed by the thalamus.

✓ **Answer C**
   – The medial wall is formed by the body of the fornix below.

**15. Lateral ventricular walls, the atrium, the FALSE answer is:**
   A. The roof is formed by the body, splenium and tapetum of the corpus callosum.
   B. The medial wall is formed by the forceps major.
   C. The lateral wall is formed by the caudate anteriorly and tapetum posteriorly.
   D. The anterior wall is formed by crus of the fornix medially and pulvinar laterally.
   E. The floor is formed by the collateral trigone.

✓ **Answer B**
   – The medial wall is formed by two roughly horizontal prominences that are located one above the other. The upper prominence, called the bulb of the corpus callosum, overlies and is formed by the large bundle of fibers called the forceps major, and the lower prominence, called the calcar avis, overlies the deepest part of the calcarine sulcus.

**16. Lateral ventricular walls, the occipital horn, the FALSE answer is:**
   A. The posterior wall is formed by the bulb of corpus callosum and calcar avis.
   B. The roof and the lateral wall are formed by the tapetum.

    C. The floor is formed by the collateral trigone.

    D. It extends posteriorly from the atrium to the occipital lobe.

    E. It varies in size from being absent to extending far posteriorly in occipital lobe.

✅ **Answer A**

    — The medial wall is formed by the bulb of the corpus callosum and the calcar avis.

❓ **17. Lateral ventricular walls, the temporal horn, the FALSE answer is:**

    A. The floor is formed by the hippocampus medially.

    B. The floor is formed by the collateral eminence laterally.

    C. The roof is formed by calcar avis.

    D. The lateral wall is formed by the tapetum.

    E. Medial wall is formed by the choroidal fissure.

✅ **Answer C**

    — The roof is formed by the inferior surface of the thalamus and the tail of the caudate medially and by the tapetum laterally.

❓ **18. The choroidal fissure, the FALSE answer is:**

    A. It is a C-shaped cleft between the fornix and the thalamus.

    B. It lies in the medial part of the body, atrium, and temporal horns.

    C. Extends from foramen of Monro around superior, inferior, and posterior surfaces of thalamus.

    D. Ends in inferior choroidal point, medial to medial geniculate body.

    E. The choroid plexus is attached along its side.

✅ **Answer D**

    — It ends in the inferior choroidal point, just behind the head of hippocampus and lateral to the lateral geniculate body.

❓ **19. The choroidal fissure boundaries, the FALSE answer is:**

    A. In atrium, by the crus of the fornix posteriorly.

    B. In atrium, by the pulvinar anteriorly.

    C. In temporal horn, by the fimbria of the fornix below.

    D. In temporal horn, by the stria terminalis and the thalamus above.

    E. In body of lateral ventricle, by the hypothalamus superiorly.

**✔ Answer E**
- In the body of lateral ventricle, by the body of the fornix superiorly, and the thalamus inferiorly.

**❓ 20. The choroidal fissure, surgical openings, the FALSE answer is:**
  A. Through atrium exposes quadrigeminal and posterior part of ambient cisterns and pineal region.
  B. Through temporal horn exposes ambient and posterior part of crural cisterns.
  C. Through the body of lateral ventricle exposes velum interpositum and the roof of the third ventricle.
  D. Through occipital horn exposes supracerebellar cistern.
  E. Through superior part of the choroidal fissure to the frontal horn.

**✔ Answer D**
- The choroidal fissure does not extend to the occipital horn.

**❓ 21. The choroid plexus, the FALSE answer is:**
  A. It extends from lateral to the third ventricles through the foramen of Monro.
  B. In the atrium it forms a prominent tuft called the glomus.
  C. Choroidal arteries arise from ICA and posterior cerebral arteries and enter through choroidal fissure.
  D. Venous drainage reaches the internal cerebral, basal, and greater veins.
  E. The choroidal fissure is the thickest part of ventricular wall.

**✔ Answer E**
- The choroidal fissure is the thinnest part of the ventricular wall, due to lack of intervening nervous tissue.

# Lateral Ventricle: Pathology: Tumors

*Abdullah H. Al Ramadan, Sarah A. Basindwah, Sadeem A. Albulaihed, Abdulaziz Y. Alahmed, Abdulaziz S. Alayyaf, Wafa M. Almuallim, and Mustafa Ismail*

**?** **1. Intraventricular tumors, the FALSE answer is:**
   A. Include both benign and malignant tumors.
   B. They are classified as extra-axial, and as intra-axial lesions.
   C. Hydrocephalus is a rare presentation.
   D. They frequently cause obstruction in cerebrospinal fluid (CSF) flow.
   E. Less than 1% of intracranial lesions.

**✓ Answer C**
   - Hydrocephalus is the main related presentation.

**?** **2. Intraventricular tumor, the FALSE answer is:**
   A. Most are benign.
   B. Most are growing slowly.
   C. Most common primary tumor is choroid plexus papilloma.
   D. Most common secondary tumor is meningioma.
   E. 80% of the ventricular tumors are confined to the ventricular cavity.

**✓ Answer E**
   - Only around 10% of these masses are confined to the ventricular system.
   - Common primary tumors are choroid plexus papillomas, ependymomas, and craniopharyngiomas. Common secondary tumors include meningiomas, gliomas, and pituitary adenomas.

**?** **3. Lateral ventricle tumor, the common immunohistochemistry marker, the FALSE answer is:**
   A. Astrocytoma (I-IV) → GFAP.
   B. Astrocytoma (I-IV) → S100.
   C. Ependymoma → GFAP.
   D. Choroid plexus tumors → CK.
   E. Choroid plexus tumors → SYN.

**✓ Answer E**
   - Common immunohistochemistry markers for choroid plexus tumors are EMA, CK, and rarely GFAP/S100.
   - SYN marker is commonly used when neurocytoma, ganglioglioma, medulloblastoma (MB), or metastatic neuroendocrine tumors are suspected.

**4. Imaging features for intraventricular tumors, the FALSE answer is:**
   A. Ependymomas showed intratumoral cyst and calcification of half of the patients.
   B. Choroid plexus papilloma bleeding can be observed frequently.
   C. MRI is useful to differentiate between Choroid Plexus Papilloma and Carcinomas.
   D. Craniopharyngioma is associated with cystic and solid component in 90%.
   E. Meningiomas are isointense in T1-weighted images.

**Answer C**
   – MRI does not readily permit differentiation of CPP and CPC. The only differentiating test is through histopathology as their radiological characteristics are quite overlapping.

**5. Choroid plexus papilloma (CPP), the FALSE answer is:**
   A. CPP is a WHO grade I intraventricular papillary neoplasm.
   B. P53 mutations are common in CPPs.
   C. Histology shows extracellular choroid plexus tissue.
   D. Immunoreactive to transthyretin, cytokeratins, and synaptophysin.
   E. Immunoreactive to focal glial fibrillary acidic protein (GFAP).

**Answer B**
   – Unlike choroid plexus carcinoma, there is no p53 mutation in CPPs.

**6. CPP, the FALSE answer is:**
   A. More common in children.
   B. Usually extend beyond the margin of the ventricles.
   C. More common in children younger than 2 years.
   D. Calcifications can be seen in CPP.
   E. CPP are usually lobulated and create small flow voids.

**Answer B**
   – CPPs are usually benign tumors that do not extend beyond the margin of the ventricles, unlike choroid plexus carcinoma.

**7. CPP, the FALSE answer is:**
   A. It is common to see the lesions in the lateral and third ventricles.
   B. Symptoms of increased Intracranial pressure (ICP) result from increased production of CSF.

C. More common in males.
D. Malignant transformation occurs in 20% of children.
E. Half of them located in the atrium.

✓ **Answer C**
 — There is no predilection to occur in either sex.

**8. CPP, the FALSE answer is:**
A. They are more common in pediatrics than adult population.
B. Patients present with symptoms of high ICP.
C. They are most commonly located in the third ventricle.
D. Homogenous enhancement on computed tomography (CT) scan.
E. On MR spectroscopy, choroid plexus papillomas tend to have an elevated myoinositol peak.

✓ **Answer C**
 — CPPs are more commonly seen in the lateral, followed by 30% in the fourth ventricle, and then the third ventricle and cerebellopontine angle.

**9. CPP, the FALSE answer is:**
A. Location is more common as lateral ventricle 50%.
B. May arise in Aicardi syndrome.
C. Presentation is with increasing head circumference or hydrocephalus.
D. Imaging appearances non-enhancing intraventricular mass.
E. Gross appearance of well-defined cauliflower-like mass.

✓ **Answer D**
 — Imaging appearances of CPP is a well-circumscribed intraventricular mass. The CPP is enhanced with gadolinium and T2 hyperintense.

**10. CPP, the FALSE answer is:**
A. Most common location in children is the left lateral ventricle.
B. Most common location in adults is lateral ventricle.
C. Enlargement of the ventricles can be due to excessive CSF production.
D. Enlargement of the ventricles can be due to tumor growth.
E. Enlargement of the ventricles can be due to CSF entrapment.

✓ **Answer B**
 — Choroid plexus papillomas in the adult population are often found at the caudal aspect of the fourth ventricle, and often calcify.

**? 11. CPP, the FALSE answer is:**
   A. 10% of CPPs occur before the age of five.
   B. On imaging, it is an avidly enhancing mass with frond-like margins.
   C. Positive immunohistochemistry for transthyretin can differentiate CPPs from other tumors.
   D. Histologically, it is marked by a proliferation of bland-looking choroid plexus epithelial cells (hyperplasia).
   E. In adult, the most common location is the fourth ventricle.

**✓ Answer A**
   ▬ More than 80% of CPPs occur before the age of five.

**? 12. CPP Histopathology, the FALSE answer is:**
   A. Presence of vascular connective tissue stroma.
   B. Increased nuclear to cytoplasmic ratios, increased mitotic figures.
   C. Has single layer of columnar epithelium.
   D. Absence of cilia.
   E. Melanotic pigmentation may be seen.

**✓ Answer B**
   ▬ Choroid plexus papillomas are WHO grade I, they show epithelial cells are more crowded and piled up than normal with mild atypia.

**? 13. CPP, the FALSE answer is:**
   A. It constitutes about 3% of all intracranial tumors in children.
   B. They exhibit columnar epithelium in papillary extensions.
   C. They are associated with Li-Fraumeni syndrome.
   D. They are vimentin, GFAP and S100 positive.
   E. Needle biopsy provides definitive diagnosis.

**✓ Answer E**
   ▬ Needle biopsy is not recommended since histologically resembles normal choroid plexus.

**? 14. CPP Histopathology, the FALSE answer is:**
   A. Exhibits a columnar epithelium with an underlaying fibrovascular network.
   B. Positive for S-100, transthyretin, and cytokeratin.
   C. Composed of papillae of squamous cells without fibrovascular cores.

D. Shows prominent papillary projections.
E. Loss of "cobble stone" surface that is seen in normal choroid plexus.

**Answer C**
– CPPs are composed of columnar epithelium with an underlying fibrovascular network, unlike papillary craniopharyngiomas which are composed of papillae of squamous cells without fibrovascular cores.

**15. Choroid Plexus Carcinoma, the FALSE answer is:**
A. They have more heterogenous appearance.
B. They tend to extend beyond ventricular margins.
C. They are rare highly malignant tumors (WHO grade III).
D. The majority of CPCs appear in age more 40 years.
E. On MR spectroscopy, choroid plexus carcinomas are more often associated with an elevated choline peak.

**Answer D**
– About 80% of carcinomas arise in children, the majority <2 years of age.

**16. CPC, the FALSE answer is:**
A. Choroid plexus carcinoma is (WHO grade I).
B. Nearly all choroid plexus carcinomas are P53 reactive.
C. Histologically, it resembles atypical teratoid/rhabdoid tumors.
D. Grossly, it shows gray-tan tumor with areas of hemorrhage and necrosis.
E. Brain invasion may be evident.

**Answer A**
– Choroid plexus carcinoma is a malignant choroid plexus neoplasm (WHO grade III).

**17. CPC, the FALSE answer is:**
A. Poor prognosis due to CSF dissemination.
B. Positive cytokeratin immunoreactivity.
C. Associated with Li- Fraumeni syndrome.
D. Histologic diagnosis requires hypercellularity only.
E. Negative EMA immunoreactivity.

**Answer D**
– Histological criteria (at least four of the following): frequent mitoses (>5 of 10 high power fields), increased cellularity, nuclear pleomorphism, blurring of papillary pattern, necrosis.

**? 18. Ependymomas, the FALSE answer is:**
   A. More common in children.
   B. Peak in the first 3 years of age.
   C. More common in males.
   D. Show perivascular pseudorosettes and ependymal rosettes.
   E. Most are classified as WHO grade I.

**✓ Answer E**
   — Most ependymomas are classified as grade II tumors according to WHO classifications.

**? 19. Subependymoma, the FALSE answer is:**
   A. It is a benign glial neoplasm (WHO grade I).
   B. Usually attached to the wall of lateral ventricles.
   C. Mean age of presentation is 60 years old.
   D. On imaging it shows small, nodular, discrete mass (<2 cm), occasional foci of calcification, hemorrhage or cystic change.
   E. Usually presents with hydrocephalus.

**✓ Answer E**
   — Subependymomas are mostly found incidentally on imaging or autopsy though very rare to present with hydrocephalus.

**? 20. Subependymoma, the FALSE answer is:**
   A. They are lobular neoplasms that are well demarcated from adjacent central nervous system (CNS) tissue.
   B. Immunohistochemistry shows GFAP-negative cell processes, positive for synaptophysin.
   C. Histologically, it shows nodular growth pattern, clusters of small bland nuclei within background of eosinophilic fibrillary processes.
   D. They show microcystic changes if found in the lateral ventricles.
   E. They have vague perivascular pseudorosette-like pattern.

**✓ Answer B**
   — Immunohistochemistry shows GFAP-positive cell processes, negative for synaptophysin.

**? 21. Subependymoma, the FALSE answer is:**
   A. They are associated with TSC-1 or TSC-2 mutations.
   B. They characteristically show no enhancement on imaging.
   C. Incidence peaks in the fifth and sixth decades of life.

**2**

D. Histologically, they show loose aggregates of tumor cell nuclei with intervening hypocellular zones.
E. May show microcystic changes on imaging.

✓ **Answer A**
- Subependymomas are not associated with *TSC-1* or *TSC-2* mutations, in contrast to the subependymal giant cell astrocytoma (SEGA) of tuberous sclerosis.

❓ **22. Subependymoma, the FALSE answer is:**
A. Small groups of cells with scant cytoplasm in a fibrillary background.
B. The classic histological picture is "islands of blue in a sea of pink.
C. They are poorly enhancing ventricular masses on MRI.
D. Typically attached to the ventricular wall.
E. Mostly affects children.

✓ **Answer E**
- It mostly affects middle aged and elderly adults, occasionally children.

❓ **23. Subependymoma, the FALSE answer is:**
A. They are slow growing with an indolent clinical course.
B. They are often discovered incidentally.
C. They can arise in the third ventricle.
D. Subtle enhancement on imaging.
E. They may arise in frontal horns of the lateral ventricles.

✓ **Answer C**
- They tend to occur in adults and most often arise in the region of the lower medulla and project into the fourth ventricle.

❓ **24. Subependymoma, the FALSE answer is:**
A. Enhancing mass with contrasted MRI.
B. Proliferation of fibrillary subependymal astrocytes.
C. Calcifications or hemorrhage may be present.
D. Well circumscribed and subependymal in location.
E. They do not fungating into the ventricle.

✓ **Answer A**
- They are non-enhancing sharply demarcated, nodular mass on imaging.

**? 25. Subependymal giant cell astrocytoma (SEGA), the FALSE answer is:**
  A. They usually arise in the fourth ventricle.
  B. They are benign intraventricular neoplasms.
  C. They present with astrocytic and neuronal features.
  D. They are associated with tuberous sclerosis complex.
  E. Usually seen in the first two decades in life.

✓ **Answer A**
  — SEGA usually appears in the lateral ventricles near the foramen of Monro, and sometimes in the third ventricles.

**? 26. SEGA, the FALSE answer is:**
  A. Astrocyte-like tumor cells can appear polygonal with glassy eo-sinophilic cytoplasm, spindled, or epithelioid cells on histology.
  B. Histologically, they are arranged in fascicles or nests separated by "gangloids."
  C. Perivascular pseudorosette-like arrangement of tumor cells is common.
  D. If mitotic figures are seen, it is an indication of poor prognosis.
  E. It has strong GFAP in subsets and is S100 reactive.

✓ **Answer D**
  — Mitotic figures may be present, but they are of no prognostic value in SEGA.

**? 27. SEGA, the FALSE answer is:**
  A. On imaging, they appear as solitary, circumscribed intensely enhancing intraventricular mass.
  B. Malignant transformation is common.
  C. Grossly, it is solid well-circumscribed mass with calcification, and spontaneous hemorrhage.
  D. If present, seizures are due to cortical hamartomas (tuber) not SEGA.
  E. Associated with facial angiofibroma.

✓ **Answer B**
  — Malignant transformation in SEGA is rare.

**? 28. SEGA, the FALSE answer is:**
  A. Can be unilateral or bilateral.
  B. Can develop from benign subependymal modules (hamartomas).

C. There is no need for surgical resection if the tumor is asymptomatic.

D. Can block CSF circulation and present with high ICP symptoms.

E. Develops in 10–15% of individuals with tuberos sclerosis complex.

✅ **Answer C**

- Earlier intervention has been proposed to avoid sequelae of hydrocephalus.

- Indications for surgery include asymptomatic SEGAs with documented tumor growth or enlargement of the ventricles is observed; symptomatic SEGA (e.g., behavioral changes, worsening of seizures, symptoms of raised ICP).

❓ **29. SEGA, the FALSE answer is:**

A. Histologically, consist of balloon cells with prominent nucleoli.

B. They are non-enhancing lesions on imaging.

C. Spontaneous malignant transformation is rare.

D. Usually present between 10 and 30 years of age.

E. They are almost exclusively seen in the setting of tuberous sclerosis.

✅ **Answer B**

- SEGAs and subependymal nodules both enhance on imaging and are difficult to differentiate in imaging and pathology.

❓ **30. SEGA, the FALSE answer is:**

A. 15% of tuberous sclerosis patients develop SEGA.

B. Tumor can grow to cause hydrocephalus and death within a year.

C. Histologically, perivascular pseudorosette formation is common.

D. GFAP and S-100 positive immunohistochemistry.

E. Tuberous sclerosis patients develop SEGAs in utero.

✅ **Answer E**

- Children with TS are seldom born with SEGAs, and commonly appears in age 8–18 months and presented later present between 10 and 30 years of age.

❓ **31. Central neurocytoma, the FALSE answer is:**

A. Predilection for the occipital horns.

B. Occur between second and fourth decades.

   C. Less than 1% of intracranial tumors.
   D. Equal distribution between both genders.
   E. It may show "fried egg" appearance in histopathological examination.

✅ **Answer A**
- Typically occur in the lateral ventricle with predilection for the left frontal horn.

❓ **32. Central neurocytoma, the FALSE answer is:**
   A. Usually located in the lateral ventricles.
   B. Homogenous vascular staining on angiography.
   C. On T1, it is isointense to the cerebral cortex and on T2 weighted and proton density images.
   D. It has got the highest oxidative metabolic rate as compared to the other tumor.
   E. It has the worst prognosis amongst intraventricular tumors.

✅ **Answer E**
- Central neurocytomas are histologically benign and have an excellent prognosis. The 5-year survival rate is 81%.

❓ **33. Central neurocytoma, the FALSE answer is:**
   A. Differentiated from other ventricular tumors by positive synaptophysin stain.
   B. They are enhancing ventricular masses on MRI.
   C. Usually originate from the atrium of the lateral ventricles.
   D. Histologically, they show a cellular tumor that has abundant cytoplasm.
   E. They are difficult to differentiate from oligodendroglioma in histology.

✅ **Answer C**
- It usually originates from the septum pellucidum, it is the most common neoplasm of septum pellucidum in young adults.

❓ **34. Central neurocytoma, the FALSE answer is:**
   A. The small astrocytic component is often GFAP positive.
   B. Synaptophysin positive.
   C. Ultrastructure of dense core with clear vessels and intermediate filaments.
   D. A sheet of cells with uniformed round nuclei.
   E. Positive immunoreactivity to Ki67.

✅ **Answer E**
- Central neurocytomas have negative immunoreactivity to Ki67, with only less than 2% positive staining.

❓ **35. Central neurocytoma, the FALSE answer is:**
A. Seen in the first decade of life.
B. CT scan shows iso- to hyperdense intraventricular lesion.
C. MRI T1 shows isointense mass.
D. On T2-weighted image, it shows iso- slightly hyperintense mass with heterogenous enhancement.
E. Lobulated mass that is adjacent to the septum pellucidum.

✅ **Answer A**
- Central neurocytomas are usually seen in the second to third decade of life.

❓ **36. High grade glioma, the FALSE answer is:**
A. The tendency for malignant transformation in astrocytoma is correlated to the patient's age.
B. HGG of the lateral ventricle may arise from the corpus callosum.
C. Has predilection for the anterior horn of the lateral ventricle.
D. More common in males.
E. The mean age at diagnosis peaks at 60 to 70 in anaplastic astrocytoma.

✅ **Answer E**
- The mean age at diagnosis peaks at 45 to 51 years in anaplastic astrocytoma and 45 to 75 in glioblastoma.

❓ **37. Interventricular meningioma, the FALSE answer is:**
A. It has male predominance.
B. It appears in mid adulthood.
C. It is the second most common tumor of the lateral ventricles.
D. It originates from arachnoid cap cells of the choroid plexus and the tela choroidea.
E. It is present as vividly enhancing mass at the trigone of the lateral ventricle.

✅ **Answer A**
- Intraventricular meningioma has a female predominance.

**38. Interventricular meningioma, the FALSE answer is:**
    A. The most common trigonal neoplasms in adults are intraventricular meningiomas.
    B. They are homogenously enhanced on MRI.
    C. 2% of all meningiomas are intraventricular.
    D. They arise from the choroid plexus stromal cells.
    E. They are the most common neoplasm in the trigone in children.

**✓ Answer E**
    – The most common trigonal mass in children is choroid plexus papilloma.

**39. Interventricular meningioma, the FALSE answer is:**
    A. The arterial supply stems from choroidal arteries.
    B. They drain to deep ventricular veins.
    C. The histopathological features always show more aggressive nature.
    D. The majority are WHO grade I.
    E. Usually become clinically evident due to obstruction of the CSF pathway.

**✓ Answer C**
    – Their histopathologic features and natural behavior are the same as those in other locations.

**40. Hemangiopericytomas, the FALSE answer is:**
    A. Can be differentiated from meningiomas by negative EMA stain.
    B. Most hemangiopericytoma are WHO grade II.
    C. Intratumoral staghorn vessels are classically seen in histology.
    D. Negative immunohistochemistry to vimentin.
    E. Homogenously enhancing on imaging resembling a meningioma.

**✓ Answer D**
    – Hemangiopericytomas have positive immunohistochemistry to vimentin, with negative EMA.

# Lateral Ventricle: Pathology: Non-Tumors

*Maliya Delawan, Younus M. Al-Khazaali,
Soubhagya R. Tripathy, Aanab O. Abdulameer,
Ali A. Dolachee, and Mohammed A. Alrawi*

**? 1. False regarding "CRASH Syndrome" Is:**
   A. Usually associated with hereditary stenosis of aqueduct of Sylvius.
   B. Includes CC (corpus callosum) hypoplasia, mental retardation, adducted thumbs, spastic paraplegia, hydrocephalus.
   C. Caused by a mutation in the (L1 Cell adhesion molecule) gene.
   D. Similarity with fetal alcohol syndrome is regional increase in cortical thickness.
   E. Also known as X-linked hydrocephalus.

**✓ Answer D**
   - Similarity with fetal alcohol syndrome and CRASH Syndrome is corpus callosal dysgenesis. Regional increase in cortical thickness is specific to fetal alcohol syndrome.

**? 2. Hemi-megalo-encephaly. False answer is:**
   A. Enlargement and cyto-architectural abnormalities of one cerebral hemisphere.
   B. Diminution of progenitor cells with failure of normal post neuro-genesis apoptosis.
   C. Also known as TORopathies-aberrant mTOR signaling.
   D. Associated with syndromes like Proteus, Klippel-Weber-Trenaunay, Epidermal Naevus, NF1, hypomelanosis of ito.
   E. Grey-white matter junction is blurred; balloon cells are identified in 50% cases.

**✓ Answer B**
   - Hemi-megalo-encephaly represents abnormally increased proliferation of progenitor cells.
   - "Lumpy-bumpy" thickened cortex on T1 scans.
   - Dystrophic calcification.
   - Major differential diagnosis of hemi-megalo-encephaly is focal malformation of cortical development.

**? 3. Periventricular nodular heterotopia. The false answer is:**
   A. Most common MCD in adults.
   B. Subependymal grey matter nodules line the lateral wall of ventricles.
   C. Brilliantly contrast enhancing.
   D. Common along temporal and occipital horns.
   E. FLNA mutation is lethal in boys—bilateral grey matter nodules, X-linked dominant form.

✅ **Answer C**
- Similar in intensity to grey matter and non-contrast enhancing.
- Associated with Dandy-Walker, Chiari 2 malformations.
- Nodules are non-calcifying (D/D-tuberous sclerosis nodules).

❓ 4. **Classic Lissencephaly. The false answer is:**
   A. Class-1 lissencephaly, four layer lissencephaly, agyria-pachygyria complex are synonymous with classic lissencephaly.
   B. Parieto-occipital agyria, with shallow sulci along fronto-temporal lobes.
   C. Miller-Dieker syndrome constitutes classic lissencephaly, frontal bossing, hypertelorism, upturned nose, small jaw, prominent upper lip, thin vermilion border.
   D. Hour-glass or Figure of eight, on NCCT scans.
   E. Band heterotopia or double cortex appearance is seen on MRI.

✅ **Answer E**
- Variant lissencephaly shows double cortex appearance—a homogenous layer of grey matter separated from ventricles and cortex by layers of normal appearing white matter.

❓ 5. **Cobblestone lissencephaly. The false answer is:**
   A. Also called—LIS2 or type-2 lissencephaly.
   B. Brain surface and ventricle surfaces are smooth.
   C. Associated with congenital muscular dystrophies.
   D. Cytomegalovirus lissencephaly shows periventricular calcifications.
   E. RELN or LIS2 gene ('Reelin' gene)—promotes neuronal migration and cell positioning.

✅ **Answer B**
- Brain surface is pebbly, not smooth as in classic lissencephaly.
- (Cobblestone complex) includes—Walker-Warburg syndrome,
- (muscle, eye, brain) disease (Fukuyama congenital muscular dystrophy).

❓ 6. **Colpocephaly. The false answer is:**
   A. It is a form of congenital non-obstructive ventriculomegaly.
   B. It is commonly associated with full or partial agenesis of the corpus callosum.

C. It is characterized by disproportionate dilation of the anterior horns.

D. The posterior to anterior ratio may aid in distinguishing colpocephaly from NPH.

E. Patients present with motor abnormalities, intellectual disability, vision problems and seizures.

**Answer C**

— It is characterized by disproportionate dilation of the posterior horns. The posterior to anterior (P/A) ratio is useful in distinguishing colpocephaly from NPH, where a ratio 23 is specific for colpocephaly. It is calculated by dividing the maximal width of the occipital horn by the maximal width of the anterior horn of the lateral ventricle.

**7. Colpocephaly. The false answer is:**

A. It is usually diagnosed in the perinatal period.

B. It is associated with the abnormal arrested migration of neuroblasts.

C. There is reduced white matter thickness in the posterior part of the centrum semiovale.

D. It can be associated with neuronal migration disorders (lissencephaly, pachygyria).

E. The only known etiology is due to chromosomal anomalies.

**Answer E**

— Various etiologies have been found to cause colpocephaly, including chromosomal anomalies such as trisomy-8 mosaicism and trisomy-9 mosaicism, intrauterine infection such as toxoplasmosis, autosomal or X-linked recessive inheritance, perinatal anoxic ischemic encephalopathy and maternal use of drugs such as corticosteroids, salbutamol and theophylline in early pregnancy. It has also been linked to periventricular leukomalacia due to the destruction of the white matter surrounding the posterior horns.

**8. Schizencephaly. The false answer is:**

A. It is characterized by an abnormal slit or cleft that connects the surface of the cerebral hemisphere to the lateral ventricle.

B. It results from abnormal neuronal migration.

C. It is classified into trans-mantle, closed lip, and open lip.

D. It can be associated with the absence of septum pellucidum.

E. It is lined by white matter.

✅ **Answer E**
- It is lined by heterotopic grey matter.

❓ **9. Schizencephaly. The false answer is:**
   A. CSF is contained in the cleft in the closed-lip subtype.
   B. It is associated with polymicrogyria, heterotopia, and arachnoid cysts.
   C. Possible etiologies include exposure to teratogens, prenatal viral infections, genetic factors, stroke in utero and young maternal age.
   D. A cleft containing CSF is present in the trans-mantle type.
   E. The open lip type typically present with more severe symptoms than the closed lip type.

✅ **Answer D**
- The cleft in the trans-mantle type does not contain CSF. The cleft only contains a column of abnormal grey matter.

❓ **10. Schizencephaly. The false answer is:**
   A. It can be unilateral or bilateral.
   B. The temporal lobe is the most commonly involved lobe.
   C. It can be associated with septo-optic dysplasia (SOD) and optic nerve hypoplasia.
   D. An area of dysplastic cortex can be situated in the contralateral hemisphere.
   E. Patients develop developmental delay, seizures, and weakness.

✅ **Answer B**
- The frontal lobe is the most commonly involved lobe.

❓ **11. Holoprosencephaly (HE). The false answer is:**
   A. The pathogenesis involves abnormal signaling between the neural crest and neural ectoderm.
   B. The majority of live births with HE is due to chromosomal abnormalities.
   C. It may be associated with DON gene-related Steinfeld syndrome.
   D. It may be associated with pregestational maternal diabetes.
   E. There is no known sex predilection.

✅ **Answer B**
- The majority of live births with HE is not attributed to chromosomal anomalies; most are due to gene mutations or deletions, environmental factors, or other unknown factors. However, a minority of cases

are due to chromosomal anomalies, of which the most common is trisomy 13, followed by trisomy 18 and triploidy.

**12. Holoprosencephaly. The false answer is:**
  A.  The cavum septi pellucidi is absent in the lobar variant.
  B.  In the semi-lobar variant, the anterior horns of lateral ventricles are absent.
  C.  The corpus callosum is absent entirely in the semi-lobar variant.
  D.  A single midline ventricle is seen in the alobar variant.
  E.  The lobar variant is characterized by non-separation of basal frontal brains.

**Answer C**
  — Only the anterior corpus callosum is absent in the semi-lobar variant. It is characterized by non-separation of frontal lobes and also present with absent anterior horns of lateral ventricles and fused deep grey nuclei.

**13. Agenesis of the corpus callosum. The false answer is:**
  A.  The lateral ventricles are often close together and oriented parallel to each other.
  B.  It is often associated with non-obstructive hydrocephalus.
  C.  Maternal phenylketonuria can be a potential risk factor.
  D.  It can be part of Apert syndrome.
  E.  They can be caused by chromosomal anomalies including deletions or duplications as well as trisomies 18 and 13.

**Answer A**
  — The lateral ventricles are often far apart and oriented parallel to each other.

**14. Agenesis of the corpus callosum. The false answer is:**
  A.  A reduction in the number of von Economo neurons can be seen.
  B.  The core syndrome consists of reduced interhemispheric transfer of sensory-motor information and developmental delay.
  C.  Patients can present with mental retardation and vision problems.
  D.  Socio-behavioral disorders may be present.
  E.  It is often accompanied by seizures.

**Answer B**
  — The core syndrome consists of reduced interhemispheric transfer of sensory-motor information, delay in cognitive processing and deficits in complex reasoning and unacquainted task performance.

**② 15. Agenesis of the corpus callosum. The false answer is:**
A. Fetal ultrasound can detect AgCC as early as the 16th week of gestation.
B. A dilated and elevated third ventricle may be seen.
C. There may be associated colpocephaly.
D. The racing car sign pertains to the appearance of lateral ventricles on sagittal sections.
E. The moose head appearance of frontal horns on coronal views is due to CCA and eversion of bilateral cingulate gyri into the frontal horns.

**✓ Answer D**
– The racing car sign pertains to the appearance of lateral ventricles on axial sections.

**② 16. Agenesis of the corpus callosum. The false answer is:**
A. There is an absent septum pellucidum.
B. The interhemispheric fissure is reduced.
C. The normal semicircular arterial loop is missing or distorted.
D. Inverted Probst bundles run parallel to and deform the medial borders of lateral ventricles.
E. The sunray appearance can be appreciated on the sagittal view.

**✓ Answer B**
– The interhemispheric fissure is widened.

**② 17. Septo-optic dysplasia. The false answer is:**
A. It is associated with younger maternal age.
B. Can be caused by mutations in HESX1 and SOX2.
C. Diagnosis is clinical and genetic evidence is not necessary for diagnosis.
D. Incidence is 1 in 10,000 live births.
E. The classical triad consists of optic nerve hypoplasia, hydrocephalus, and midline brain defects.

**✓ Answer E**
– The classical triad consists of optic nerve hypoplasia, pituitary hormone abnormalities, and midline brain defects. Classically, the diagnosis can be made with the presence of two or more features of the triad.

**? 18. Hydranencephaly. The false answer is:**
  A. Patients can have intact primitive reflexes at birth.
  B. A common cause is infarction secondary to bilateral internal carotid artery (ICA) obstruction in the neurogenic phase.
  C. It occurs during the second trimester.
  D. It is characterized by the absence of the cerebral hemispheres and basal ganglia with intact brainstem and meninges.
  E. It can be associated with Fowler's syndrome.

**✓ Answer D**
  – It is characterized by the absence of the cerebral hemispheres with preserved basal ganglia, brainstem, and meninges.

**? 19. Hydranencephaly. The false answer is:**
  A. It can be associated with Trisomy 13.
  B. Most patients die after a few weeks from birth.
  C. It is associated with renal and heart disease.
  D. There is no sex predilection.
  E. Infants may appear completely normal at birth and present a few weeks later.

**✓ Answer B**
  – Most patients die before birth.

**? 20. Porencephalic cyst. The false answer is:**
  A. It is a CS-filled cavity in the brain parenchyma.
  B. It is most commonly caused by ischemia.
  C. It is lined by grey matter.
  D. It extends from the cortex to ventricles.
  E. Can be divided into developmental, congenital, internal and external types.

**✓ Answer C**
  – It is lined by gliotic white matter.

**? 21. Intraventricular simple cysts. The false answer is:**
  A. They are usually asymptomatic.
  B. They can be lined by cuboidal choroidal cells.
  C. They can be lined by ependymal tissue.
  D. They can be lined by arachnoid epithelial tissue.
  E. They typically have a thick wall.

✅ **Answer E**
  - They typically have very thin or imperceptible walls.

❓ **22. Intraventricular simple cysts. The false answer is:**
  A. They are most commonly found in the trigone of the lateral ventricles.
  B. On MRI, they are demonstrating enhancement and restricted diffusion.
  C. They are associated with obstructive hydrocephalus.
  D. They are CSF-filled cysts.
  E. Some cases are thought to arise from the vascular mesenchyme.

✅ **Answer B**
  - On MRI, they are demonstrating no enhancement or restricted diffusion.

❓ **23. Fetal ventriculomegaly. The false answer is:**
  A. The diagnosis is made when the atrial diameter of the ventricle is more than 5 mm on ultrasound.
  B. Viral causes include Toxoplasma gondii, rubella, cytomegalovirus, and herpes simplex.
  C. It can be due to inherited X-linked mutations.
  D. It can be due to chromosomal abnormalities.
  E. The preferred time for measuring the atrial diameter is during the second and third trimesters.

✅ **Answer A**
  - The diagnosis is made when the arterial diameter of the ventricle is more than 10 mm on ultrasound.

❓ **24. Aqueductal stenosis. The false answer is:**
  A. In infants, a common presentation is the setting sun phenomenon.
  B. Bickers-Adams-Edwards syndrome is an autosomal dominant disorder that presents with aqueduct stenosis.
  C. It can be a complication of infectious ventriculitis.
  D. It can be due to subarachnoid hemorrhage (SAH).
  E. In infants, there can be macrocephaly.

✅ **Answer B**
  - Bickers-Adams-Edwards syndrome is an X-linked disorder that presents with aqueduct stenosis. It is also associated with profound intel-

lectual disability, adducted thumb, corpus callosum agenesis, absence of the medullary pyramids, and corticospinal tracts agenesis.

**25. Chiari II malformation. The false answer is:**
   A. It is characterized by beaked midbrain, downward displacement of the tonsils and cerebellar vermis, and spinal myelomeningocele.
   B. It is frequently associated with hydrocephalus.
   C. It is associated with cerebellar dysplasia.
   D. In adults, it typically presents progressive hydrocephalus.
   E. Neonates can present with apneic spells.

**Answer D**
   — In adults, it typically presents as scoliosis or syrinx. Children typically present with progressive hydrocephalus.

**26. Chiari II malformation. The false answer is:**
   A. On antenatal ultrasound, the lemon sign refers to the appearance of the indentation in the occipital bone resembling a lemon.
   B. On antenatal ultrasound, the banana sign describes the cerebellum wrapping around the brain stem.
   C. The tectal plate appears beaked.
   D. There is no sex predilection.
   E. It can be associated with sensorineural hearing loss.

**Answer A**
   — On antenatal ultrasound, the lemon sign refers to the appearance of the indentation in the frontal bone resembling a lemon.

**27. Ventriculitis. The false answer is:**
   A. Presenting features include signs of meningism.
   B. Patients can present with features of obstructive hydrocephalus.
   C. An increase in CSF lactate, procalcitonin and lysozymes suggest that the infection is viral rather than bacterial.
   D. Multiloculated hydrocephalus, which is more common with bacteria than viral ventriculitis.
   E. External ventricular drains (EVDs) or intraventricular shunts are potential risk factors.

**Answer C**
   — An increase in CSF lactate, procalcitonin, and lysozymes suggest that the infection is bacterial rather than viral.

**? 28. Ventriculitis. The false answer is:**
A. For catheter-related ventriculitis, the preferred antibiotics are vancomycin and an anti-pseudomonal beta-lactam.
B. Ultrasound done for neonates shows increased thickness, irregularity, and echogenicity of ependyma.
C. MRI shows ventricular debris hypointense to CSF on T1 and hyperintense to CSF on T2.
D. The ventricular debris may show restricted diffusion.
E. Periventricular low density may be seen on non-contrast CT.

**✓ Answer C**
— MRI shows ventricular debris hyperintense to CSF on T1 and hypointense to CSF on T2.

**? 29. Intraventricular cavernoma. The false answer is:**
A. There is no surrounding edema on T2 MRI sequences.
B. Partial calcification can be seen on CT.
C. They exhibit mixed signal intensity on MRI.
D. They are moderately hypodense on CT.
E. CT demonstrates mild contrast enhancement.

**✓ Answer D**
— They are moderately hyperdense on CT.

**? 30. Intraventricular arteriovenous malformation. The false answer is:**
A. They exhibit high intensity with a signal void on T2-weighted MRIs.
B. Most intraventricular AVMs manifest with spontaneous intracranial hemorrhage.
C. Brain digital subtraction angiography is the most reliable imaging modality for diagnosis.
D. They have high attenuation on CT.
E. They exhibit high intensity on T1-weighted MRIs.

**✓ Answer E**
— They exhibit low intensity on T1-weighted MRIs.

**? 31. Intraventricular cavernoma. The false answer is:**
A. Are composed of vessels lined by a single layer of endothelial cells and collagen fibers.
B. They contain intervening nervous tissue.
C. It consists of vessels of various sizes.

D.  They tend to grow rapidly.
E.  They tend to result in hemorrhage.

✅ **Answer B**

➖ They do not contain intervening nervous tissue.

❓ **32. Intraventricular venous malformation. The false answer is:**
A.  They are usually clinically silent.
B.  They are hypointense on T1-weighted MRIs.
C.  They can have high or low signal intensity on T2-weighted MRIs.
D.  It should be surgically resected.
E.  They are intensely enhancing on T1-weighted MRIs.

✅ **Answer D**

➖ Surgical resection is contraindicated because the venous malformation may provide important drainage for the normal brain.

❓ **33. Subependymal hamartoma. The false answer is:**
A.  They are associated with obstructive hydrocephalus.
B.  They typically measure more than 1 cm.
C.  There is variable contrast enhancement on CT.
D.  They have variable signs on T1- and T2-weighted MRIs.
E.  Calcification is commonly seen on CT.

✅ **Answer B**

➖ They typically measure less than 1 cm.

# Lateral Ventricle: Surgery

*Viraat Harsh, Manoj Kumar, Anil Kumar, Margarida S. Conceicao, Osama M. Al-Awadi, Mustafa Ismail, and Samer S. Hoz*


H. R. Salih et al. (eds.), *Cerebral Ventricles*, https://doi.org/10.1007/978-3-031-42342-0_4

**? 1. Surgical anatomy of the lateral ventricle, the false answer is:**
   A.  The caudate nucleus forms the roof of the temporal horn.
   B.  The caudate nucleus forms the medial wall of the frontal horn.
   C.  The medial wall of the frontal horn is formed by septum pellucidum.
   D.  The roof of frontal horn is formed by the genu of the corpus callosum.
   E.  The floor of frontal horn is formed by the rostrum of the corpus callosum.

**✓ Answer B**
   — Caudate nucleus forms the lateral wall of the frontal horn.

**? 2. Lateral ventricle tumors, the false answer is:**
   A. They may originate from periventricular white matter extending into the ventricle.
   B. They may originate from the ependyma.
   C. They may originate from ectopic tissue trapped within the ventricle.
   D. The main feeder vessels are anterior choroidal and posterior hippocampal
        arteries.
   E. They may originate from the choroid plexus.

**✓ Answer D**
   — Main feeder vessels are anterior choroidal and posterior lateral choroidal
   — arteries.

**? 3. Regarding surgery of lateral ventricle tumor, the false answer is:**
   A. The primary goal of treatment is total excision.
   B. In case of CSF pathway obstruction an Endoscopic Third Ventriculostomy (ETV), EVD, or VP shunt may be applied first.
   C. Emergency surgery is indicated in acute obstructive hydrocephalus or acute intratumoral hemorrhage.
   D. Endoscopic assistance is an important adjunct to microneurosurgery of lateral ventricle tumors.
   E. There is no role of bilateral EVD.

**✓ Answer E**
   — Bilateral EVD in cases of obstruction at the level of the foramen of Monro may be applied just before surgery.

**? 4. Known approaches for lateral ventricle, the false answer is:**
   A. Anterior interhemispheric transcallosal approach.
   B. Anterior transcortical approach.
   C. Transtemporal approach.
   D. Intraparietal sulcus approach.
   E. Inferior frontal gyrus approach.

**✓ Answer E**
   - The inferior frontal gyrus approach is not used for lateral ventricle tumors.
   - Superior and middle frontal gyrus approaches have been defined.
   - The inferior temporal gyrus approach is used for surgery of lateral ventricle tumors.

**? 5. Surgery of tumors in atrium of lateral ventricle, the false answer is:**
   A. The atrium can be approached by the posterior trans-sulcal approach and posterior trans-callosal approach.
   B. Optic radiations define the lateral boundary of the atrium.
   C. Inside the atrium, the choroid plexus is found laterally.
   D. Tumor compressing the lateral wall of the atrium should be debulked piecemeal before separating it from the ependymal surface.
   E. A transtemporal approach may be used.

**✓ Answer C**
   - Inside the atrium, the choroid plexus is found medially.

**? 6. Interhemispheric approaches, the false answer is:**
   A. Risk of injury to the bridging veins or the posterior occipital vein.
   B. Injury to bridging veins anterior to the coronal suture may lead to serious complications.
   C. Transient mutism is a rare complication.
   D. In the parasplenial route, the main risk is visual field deficit.
   E. Neuronavigation is especially helpful in the parasplenial route.

**✓ Answer B**
   - Bridging veins anterior to the coronal suture may be sacrificed without serious complication.
   - Transient mutism is a rare complication associated with bilateral cingulate gyrus retraction.
   - Visual field deficit in parasplenial route is caused by retraction over the calcarine region and the internal occipital vein.

- In the parasplenial route with patients placed in a three-quarter prone position and in those with awkward tumors with difficult anatomic views neuronavigation is especially helpful.

**7. Anterior interhemispheric transcallosal approach, the false answer is:**
   A. It is also known as the frontal interhemispheric approach.
   B. A preoperative venogram helps to avoid bridging veins draining into the superior sagittal sinus (SSS).
   C. Self-retaining retractors provide the best retraction.
   D. Two to three large-tailed cotton sponges placed into the trigone help prevent blood and tumor particles from seeping into the occipital and temporal horns.
   E. Copious irrigation facilitates the removal of blood and debris at the end of the tumor excision.

**Answer C**
- Sequentially increasing sizes of soft cotton balls placed at anterior and posterior ends of dissection are preferred over self-retaining retractors.

**8. Anterior interhemispheric transcallosal approach, the false answer is:**
   A. When the interhemispheric approach is used, the pericallosal and callosomarginal arteries, as well as veins draining toward the SSS, must be preserved.
   B. It gives excellent exposure to the frontal horn.
   C. It may be used to approach ventricular trigone tumors.
   D. It can be used to approach the anterior part of the third ventricle also.
   E. It provides good exposure to the body of the lateral ventricle.

**Answer C**
- The posterior interhemispheric transcallosal approach is used for trigone tumors.

**9. Regarding the anterior interhemispheric transcallosal approach, the false answer is:**
   A. Brain relaxation can be achieved by osmotherapy.
   B. Release of CSF may be required from sulcal and callosal cisterns.
   C. Opening in the corpus callosum should be 1–1.5 cm.
   D. Dorsal mesencephalic cisterns can be opened to release CSF.

    E. In the dominant hemispheric ventricle, the contralateral transcallosal approach may be implemented to avoid related speech deficits.

✅ **Answer D**

- Dorsal mesencephalic cisterns can be opened to release CSF in the posterior interhemispheric transcallosal approach.

❓ 10. **Following steps help in decreasing the pressure intraoperatively in anterior interhemispheric transcallosal approach, the false answer is:**
    A. While dissection release of CSF from lamina terminalis cistern.
    B. Entry into the lateral ventricle after linear callosotomy and consequential CSF egress.
    C. Incision through the septum pellucidum to release CSF from the contralateral lateral ventricle and further access to both foramina of Monro.
    D. Osmotherapy.
    E. Lumbar puncture.

✅ **Answer A**

- While dissection release of CSF from the dorsal sulcal cistern and callosal cistern help to decrease the ICP.

❓ 11. **Regarding the anterior interhemispheric approach, the false answer is:**
    A. The thalamostriate vein is an important guide in the presence of a large tumor with a severely distorted ventricle.
    B. The thalamostriate vein is located left side of the plexus in the right lateral ventricle and the right side of the plexus in the left lateral ventricle.
    C. Internal cerebral vein helps in the localization of the fornix and thalamus.
    D. Complications are seizures, sensory-motor deficit, visual and cognitive impairment.
    E. There is a risk of damage to the genu of the internal capsule.

✅ **Answer B**

- The thalamostriate vein is located right side of the plexus in the right lateral ventricle and left side of the plexus in the left lateral ventricle.
- Tracing the course of the choroid plexus and vein leads to the internal cerebral vein which helps in the localization of the fornix and thalamus.

**12. Complications of anterior interhemispheric transcallosal approach, the false answer is:**
   A. Hemiparesis.
   B. Alexia with agraphia.
   C. Aphasia and mutism.
   D. Memory deficit.
   E. Astereognosis.

**Answer B**
   — Alexia without agraphia is a known complication of the anterior transcallosal approach to the lateral ventricle.

**13. For the posterior transcallosal approach, the false answer is:**
   A. Used for the tumor of the trigone and posterior body of lateral ventricle.
   B. Best for medially positioned tumors with blood supply primarily from the posterior choroidal arteries.
   C. The transcallosal approach should be avoided in patients with preexisting hemianopsia.
   D. Injury to the splenium may cause memory deficits.
   E. Can be done both in lateral decubitus and prone positions.

**Answer D**
   — Injury to the splenium may cause alexia and verbal-visual disconnection.

**14. Posterior transcallosal approach for ventricular trigone lesions, the false answer is:**
   A. Indication—recommended for lesions that extend superiorly from the trigone or involve the splenium of the corpus callosum.
   B. It is suitable for large tumors.
   C. Incision through the posterior part of the cingulate gyrus, which in turn transects the lateral part of the splenium before entering the trigone.
   D. It was originally described by Kempe and Blaylock.
   E. It is associated with an auditory or visual disconnection syndrome resulting from the posterior transection of the corpus callosum but no injury to optic radiations.

**Answer B**
   — Not suitable for large tumors as the tumor itself may prevent the hemispheric retraction required to achieve tumor removal and choroidal vascular control.

**15. For the posterior interhemispheric transcingular approach, the false answer is:**
A. It provides access to the atrium and posterior horn of the lateral ventricle.
B. Dissection of precuneus or posterior corpus callosum and cingulate gyrus is required.
C. The position is prone with craniotomy involving superior and inferior to the lambdoid suture, including the midline.
D. The posterior interhemispheric fissure is opened widely to minimize retraction.
E. The advantages of this approach are the wide surgical corridor and angle of viewing.

**Answer E**
- Narrow surgical corridor and difficult angle of viewing are disadvantages of the posterior interhemispheric transcingular approach.

**16. Regarding transcortical approaches, the false answer is:**
A. Complications are seizures, hemiparesis, memory loss, confusion, intracerebral hemorrhage (ICH), postoperative hydrocephalus, and mutism.
B. In hydrocephalic patients, the trajectory obtained from the exposure is not optimum after decompression of the ventricles by CSF drainage.
C. There is a higher risk of postoperative seizures than after transcallosal procedures.
D. Auditory complications are not seen.
E. MR venography is essential to avoid bridging veins.

**Answer E**
- Bridging veins are not encountered in transcortical approaches.
- The reported risk of postoperative seizures after transcortical approaches ranges from 30 to 70%, whereas after transcallosal procedures, the reported risk is 0–10%.

**17. Regarding the anterior transcortical approach, the false answer is:**
A. Provides excellent access to the frontal horns, the body of the lateral ventricle, and the third ventricle.
B. Suitable for the patient with large ventricle and pathology of non-dominant hemisphere.

C. It is approached through the middle frontal gyrus.
D. During dissection preservation of bridging veins is a concern.
E. The disadvantage is that commissural, projection, and association fibers need to be dissected.

**✓ Answer D**
- The advantage of the anterior transcortical approach is that bridging veins are not a concern.

**? 18. Anterior transcortical approach, the false answer is:**
A. More difficult to approach the lateral ventricle on the opposite side.
B. Higher incidence of postoperative seizures than in the transcallosal approach.
C. The transient postoperative deficit in attention can result from a middle frontal gyrus incision.
D. The incision in the supplemental motor cortex or premotor areas in either hemisphere can cause transient hemiparesis.
E. May be used to access tumors located in the midventricular body.

**✓ Answer E**
- Contraindicated for tumor located in the midventricular body because that would require making a lesion in the motor and sensory cortex.

**? 19. Posterior trans-sulcal approach, the false answer is:**
A. It is a preferred route to the body of the lateral ventricle.
B. Operative position three-quarter prone with the parietal region at the highest point.
C. Preoperative MR venogram or cerebral angiogram is needed to determine the position of major draining veins.
D. Dissection in the cortex proceeds along the inter-parietal sulcus.
E. Manipulation of the lateral wall of the atrium predispose to injury of the optic radiation.

**✓ Answer A**
- The posterior trans-sulcal approach is a preferred route to atrium of lateral ventricle.

**? 20. Transtemporal approach to lateral ventricle, the false answer is:**
A. Direct access to the temporal horn and atrium of lateral ventricle through a short trajectory.
B. Ventricular cavity can be exposed through the anterior part of middle temporal gyrus or trans-sulcal route.

    C. Advantage of this approach is early identification and management of anterior choroidal artery.
    D. Disadvantages of this approach are visual field defect and quadrantanopia.
    E. Aphasia can occur when working with dominant hemisphere.

✅ **Answer B**
- Ventricular cavity can be exposed through the posterior part of middle temporal gyrus or trans-sulcal route.

❓ **21. Occipitotemporal sulcus approach, the false answer is:**
    A. It is used for lesions of the posterior part of temporal horn.
    B. Occipitotemporal sulcus can be opened with subtemporal craniotomy.
    C. Lumbar drain may help release CSF for brain retraction before opening the dura.
    D. Risk of postoperative visual deficit is involved.
    E. It provides good posterior lateral choroidal vascular control.

✅ **Answer E**
- The posterior lateral choroidal vessels are encountered in posterior interhemispheric transcallosal approach.

❓ **22. Transcortical transparietal approach through the P1 gyrus, the false answer is:**
    A. Used to remove tumors of the trigonal region or for tumors of the posterior body of the lateral ventricle.
    B. Not suited for large masses and for tumors with a superior tumoral development or those located in the dominant hemisphere.
    C. Less chance of visual and language deficits as compared to the middle temporal gyrus approach.
    D. A disadvantage is that it does not allow anterior choroidal vascular control before tumor excision.
    E. A complication is a subdural hygroma/hematoma formation after large tumor excision.

✅ **Answer B**
- The transcortical transparietal approach is suited for large masses and for tumors with a superior tumoral development or those located in the dominant hemisphere.
- Removal of a large tumor may cause the sinking of the cortex away from the cranial topography, leading to subdural hygroma/hematoma formation.

**23. Superior parietal occipital approach for ventricular trigone lesions, the false answer is:**
   A. The incision is given along the inferior parietal gyrus.
   B. It allows access to both medial and lateral regions of the trigone.
   C. Incision- 1 cm posterior to the postcentral fissure, extending 4 to 5 cm posteriorly.
   D. Complications are apraxia, acalculia, visual field deficits, and homonymous hemianopsia.
   E. Removal of a large tumor may cause the sinking of the cortex away from the cranial topography, leading to subdural hygroma/hematoma formation.

**Answer A**
   - The incision is given along the superior parietal gyrus.
   - The incision should be made 1 cm posterior to the postcentral fissure and should extend 4 to 5 cm posteriorly, as far as the parieto-occipital sulcus.

**24. Regarding the posterior interhemispheric parieto-occipital approach to the ventricular trigone, the false answer is:**
   A. Allows preservation of precuneus gyrus.
   B. Described by Yasargil.
   C. Preferred position—sitting.
   D. Central enucleation of the tumor followed by peripheral dissection of the capsule away from the ventricular wall is performed.
   E. Ideal for lesions extending from the occipital horn to the ventricular trigone.

**Answer A**
   - The posterior interhemispheric parieto-occipital approach to the ventricular trigone requires resection of the precuneus gyrus.

**25. Regarding surgery for lesions of the temporal horn, the false answer is:**
   A. A lateral transcortical approach through the temporal or parietal gyrus is performed.
   B. Pterional transsylvian transcisternal anteromedial trajectory forms the ideal approach.
   C. Dissection between the polar and anterior temporal arteries decreases the morbidity.

D. Proximal transsylvian trans-amygdalar trajectory damages the anterior loop of Meyer's optic radiation fibers.
E. MRI-DTI helps in identifying the tracts which may be saved/sacrificed.

✓ **Answer D**
- Proximal transsylvian trans-amygdalar trajectory allows entrance into the temporal horn while minimizing the risk to the anterior loop of Meyer's optic radiation fibers as long as rigid retraction is not applied to the temporal lobe.

? **26. Complications of lateral ventricle surgery, the false answer is:**
   A. Visual field deficit.
   B. Hemiparesis and aphasia.
   C. Subdural hematoma.
   D. Auditory deficit.
   E. Persistent hydrocephalus.

✓ **Answer D**
- Auditory deficits are not seen after lateral ventricle surgery.

? **27. Regarding endoscopic excision of intraventricular tumors without hydrocephalus, the false answer is:**
   A. The head is elevated to approximately 30° above the horizontal to minimize CSF egress from the endoscope.
   B. The most direct intraventricular, linear route to the target is selected, directed by stereotaxy.
   C. In case of significant ventricular asymmetry, preference is to approach from the side having the greater ventricular size.
   D. Once inside the ventricle, gradual insufflation of the ventricular system with about 50 mL of lactated Ringer's solution helps.
   E. Hemodynamic monitoring helps in avoiding over-insufflation.

✓ **Answer D**
- Once inside the ventricle, gradual insufflation of the ventricular system with about 10 mL of lactated Ringer's solution helps.
- During insufflation, monitoring of ICP is done by looking for any relative bradycardia (by hemodynamic monitoring).
- Over-insufflation is avoided by using a separate purge channel or keeping the sheath diameter greater than the endoscope diameter thus maintaining an ICP that cannot exceed the pressure of a fluid column equal to the length of the endoscope.

# Third Ventricle: Anatomy

*Hayder R. Salih, Margarida S. Conceicao, Mostafa H. Algabri, Mahmood F. Alzaidy, Ibrahim A. Izzet, Zaid Aljuboori, and Samer S. Hoz*


H. R. Salih et al. (eds.), *Cerebral Ventricles*, https://doi.org/10.1007/978-3-031-42342-0_5

**? 1.  Third ventricle location, the FALSE answer is:**
  A.  It is below the corpus callosum.
  B.  Above the sella turcica, pituitary gland, and midbrain.
  C.  Between the two halves of the thalamus.
  D.  It is related to the circle of Willis and the great vein of Galen.
  E.  Between the body of the lateral ventricles.

**✓ Answer E**
  − The third ventricle is located below the lateral ventricles.

**? 2.  Complications related to third ventricular wall manipulation, the FALSE answer is:**
  A.  Disturbances in the temperature control.
  B.  Disturbances in the hypophyseal secretion.
  C.  Visual loss.
  D.  Sympathetic dysfunction.
  E.  Memory loss.

**✓ Answer D**
  − Manipulation of the walls of the third ventricle may cause hypothalamic dysfunction.

**? 3.  Symptoms correlated to the third ventricle's wall manipulation, the FALSE answer is:**
  A.  Disturbances of consciousness.
  B.  Loss of temperature control.
  C.  Visual loss due to damage of the optic chiasm and tracts.
  D.  Memory loss due to injury to the columns of the fornix.
  E.  Auditory system dysfunction.

**✓ Answer E**
  − Disturbances of consciousness, temperature control, respiration, and hypophyseal secretion, visual loss due to damage of the optic chiasm and tracts, and memory loss due to injury to the columns of the fornix in the walls of the third ventricle.

**? 4.  The third ventricle roof anatomy, the FALSE answer is:**
  A.  It forms a gentle upward arch.
  B.  It has three layers.
  C.  The choroidal fissure is located in its lateral margin.

   D. The upper layer of the anterior part is formed by the body of the fornix.

   E. Septum pellucidum is attached to the upper surface of the body of the fornix.

✅ **Answer B**

— The roof has four layers: one neural layer formed by the fornix, two layers made by tela choroidea, and a layer of blood vessels between the sheets of tela choroidea.

❓ 5. **Tela choroidea of the third ventricle, the FALSE answer is:**
   A. It has three layers in the roof of the third ventricle.
   B. It is situated below the fornix.
   C. It consists of thin, semi-opaque membranes.
   D. It is derived from the pia mater.
   E. The internal cerebral vein runs between its layers.

✅ **Answer A**

— The tela choroidea forms two layers in the roof of the third ventricle.

❓ 6. **The roof of the third ventricle, the FALSE answer is:**
   A. Its vascular layer is located below the two layers of the tela choroidea.
   B. Its vascular layer contains the medial posterior choroidal artery.
   C. The internal cerebral veins are found in the vascular layer.
   D. Parallel strands of choroid plexus project from the inferior layer of tela choroidea.
   E. It has three layers, two of them formed by tela choroidea and one by the fornix.

✅ **Answer A**

— The vascular layer is located between the two layers of the tela choroidea.

❓ 7. **Regarding the anatomy of the third ventricle, the FALSE answer is:**
   A. Velum interpositum lay between the layers of tela choroidea in its roof.
   B. Both layers of tela choroidea are attached to the lower surface of the fornix.
   C. Striae medullaris extends from the foramen of Monro to the habenular commissure.

    D. Suprapineal recess is located between layers of the tela choroidea and pineal body.

    E. Parallel strands of the choroid plexus in its roof are attached to the lower layer of the tela choroidea.

✅ **Answer B**

— The lower wall has an anterior part that is attached to the small ridges on the free edge of the fiber tracts, called the striae medullaris thalami.

❓ 8. **The internal cerebral veins, the FALSE answer is:**

    A. They drain the frontal horn and body and run in the velum interpositum.

    B. They are formed by veins that drain the frontal horn.

    C. They arise just anterior to the foramen of Monro.

    D. They exit the velum interpositum above the pineal body.

    E. They unite to form the greater vein of Galen in the quadrigeminal cistern.

✅ **Answer C**

— The internal cerebral veins arise in the anterior part of the velum interpositum, just behind the foramen of Monro.

❓ 9. **The velum interpositum, the FALSE answer is:**

    A. It is usually a closed space.

    B. Its major opening is situated between the splenium and pineal body.

    C. It tapers to a narrow apex behind the foramen of Monro.

    D. It communicates with the quadrigeminal cistern forming cisterna velum interpositum.

    E. Cavum vergae is a space above the velum interpositum.

✅ **Answer B**

— The velum interpositum is usually a closed space that tapers to a narrow apex just behind the foramen of Monro, but it infrequently has an opening situated between the splenium and pineal body.

❓ 10. **The floor of the third ventricle, the FALSE answer is:**

    A. It extends from the optic chiasm to the aqueduct of Sylvius.

    B. Its anterior half is formed by diencephalic structures.

    C. The whole floor is formed by mesencephalic structures.

    D. The inferior surface of the chiasm forms its anterior part.

    E. The superior surface of chiasm forms the lower part of the anterior wall.

✅ **Answer C**
- The anterior half of the floor is formed by diencephalic structures, and the posterior half is formed by mesencephalic structures.

❓ **11. The third ventricle region, the FALSE answer is:**
   A. Part of the tegmentum forms the most posterior part of the floor.
   B. The optic tracts course its floor toward the lateral margins of the midbrain.
   C. Mammillary bodies are located in space limited posteriorly by optic tracts.
   D. The hypothalamic infundibulum is located between the optic chiasm and the tuber cinereum.
   E. Axons of the infundibulum extend to the posterior hypophysis.

✅ **Answer C**
- The infundibulum, tuber cinereum, mammillary bodies, and posterior perforated substance are located in the space limited anteriorly and laterally by the optic chiasm and tracts and posteriorly by the cerebral peduncles.

❓ **12. Tuber cinereum, the FALSE answer is:**
   A. It is a prominent mass of hypothalamic gray matter.
   B. It merges anteriorly into the infundibulum.
   C. The median eminence is formed by the tuber cinereum.
   D. The mammillary bodies are located anterior to the tuber cinereum.
   E. The tuber cinereum around the base of the infundibulum is raised.

✅ **Answer D**
- The mammillary bodies form paired, rounded prominences posterior to the tuber cinereum.

❓ **13. The third ventricle floor, the FALSE answer is:**
   A. The posterior perforated substance is a depressed, punctuated area of gray matter.
   B. Its posterior part extends posteriorly and superiorly to the medial part of cerebral peduncles.
   C. Mammillary bodies on their inner surfaces are posterior to the infundibular recess.
   D. Its part between the mammillary bodies and the aqueduct of Sylvius has a smooth surface.
   E. Its smooth surface lies on both sides of the posterior perforated substance.

✅ **Answer E**

- The part of the floor between the mammillary bodies and the aqueduct of Sylvius has a smooth surface that is concave from side to side. This smooth surface lies above the posterior perforated substance anteriorly and the medial part of the cerebral peduncles and the tegmentum of the midbrain posteriorly.

❓ **14. The anterior wall of the third ventricle, the FALSE answer is:**
   A. It extends from the foramina of Monro above to the optic chiasm below.
   B. Its upper third is hidden by the rostrum of the corpus callosum.
   C. Its visible part on the surface is formed by optic chiasm and lamina terminalis.
   D. The lamina terminalis is an arachnoid membrane.
   E. The lamina terminalis extends between the optic chiasm and the rostrum of the corpus callosum.

✅ **Answer D**

- The lamina terminalis is a thin sheet of gray matter and pia mater that attaches to the upper surface of the chiasm and stretches upward to fill the interval between the optic chiasm and the rostrum of the corpus callosum.

❓ **15. The anterior wall of the third ventricle compartments, the FALSE answer is:**
   A. The fornix is superior to the foramina of Monro.
   B. The lamina terminalis is superior to the anterior commissure.
   C. The foramina of Monro is superior to the anterior commissure.
   D. The optic recess is superior to the chiasm.
   E. The lamina terminalis is superior to the optic recess.

✅ **Answer B**

- When viewed from within, the boundaries of the anterior wall are formed, from superior to inferior, by the columns of the fornix, foramina of Monro, anterior commissure, lamina terminalis, optic recess, and optic chiasm.

❓ **16. The foramina of Monro, the FALSE answer is:**
   A. On each side, it is located at the junction of the roof and the anterior wall.
   B. In the lateral ventricle, it opens between the fornix and the thalamus.

C. It is bounded anteriorly by the posterior pole of the thalamus.
D. Its size and shape depend on the size of the ventricles.
E. It extends below the fornix into the third ventricle as a single channel.

✓ **Answer C**
- The foramen of Monro is bounded anteriorly by the junction of the body and the columns of the fornix and posteriorly by the anterior pole of the thalamus.

? **17. The structures that pass through the foramen of Monro, the FALSE answer is:**
A. The choroid plexus.
B. The medial posterior choroidal arteries.
C. Thalamostriate vein.
D. Posterior choroidal vein.
E. Septal veins.

✓ **Answer D**
- The structures that pass through the foramen are the choroid plexus, the distal branches of the medial posterior choroidal arteries, and the thalamostriate, superior choroidal, and septal veins.

? **18. The posterior wall of the third ventricle, the FALSE answer is:**
A. It extends from the suprapineal recess above to the aqueduct of Sylvius below.
B. The pineal body is one of its components.
C. The shape of the orifice of the aqueduct of Sylvius is triangular with the base of the triangle is on the midbrain.
D. Suprapineal recess projects between the pineal gland and tela choroidea.
E. The stalk of the pineal gland has an upper and a lower lamina.

✓ **Answer C**
- The shape of the orifice of the aqueduct of Sylvius is triangular. The base of the triangle is on the posterior commissure and the other two limbs are formed by the central gray matter of the midbrain.

? **19. The Pineal gland, the FALSE answer is:**
A. It projects posteriorly into the quadrigeminal cisterns.
B. It is concealed by the quadrigeminal plate posteriorly.
C. It is concealed by the splenium of the corpus callosum above.

D. It is concealed by the thalamus laterally.
E. It is concealed by the vermis of the cerebellum inferiorly.

✓ **Answer B**
- The pineal gland is concealed by the vermis and quadrigeminal plate inferiorly.

? **20. The lateral wall of the third ventricle, the FALSE answer is:**
A. It is not visible on the external surface of the brain.
B. It is formed by the thalamus superiorly.
C. It is formed by the hypothalamus superiorly.
D. Its surfaces are connected by masa intermedia.
E. The inferior limit of the upper surfaces is the striae medullaris thalami.

✓ **Answer C**
- The lateral wall of the third ventricle is formed by the hypothalamus antero-inferiorly.
- The hypothalamic and thalamic surfaces of the lateral wall of the third ventricle are separated by the hypothalamic sulcus, which is a groove that is often ill-defined and extends from the foramen of Monro to the aqueduct of Sylvius.

? **21. The lateral wall of the third ventricle, the FALSE answer is:**
A. Hypothalamic and thalamic surfaces are separated by the hypothalamic sulcus.
B. The superior limit of the thalamic surface is marked by the striae medullaris.
C. The striae medullaris extends forward from the habenulae.
D. The habenulae are found on the surfaces of the thalamus.
E. The habenulae are connected by posterior perforated substance.

✓ **Answer E**
- The habenulae are connected across the midline in the rostral stalk of the pineal gland by the habenular commissure.

? **22. The tentorial incisura, the FALSE answer is:**
A. The third ventricle is situated below the tentorial incisura.
B. Its base is located in the dorsum sellae.
C. The body of the lateral ventricle is directly above its central part.
D. The midbrain is situated in the center of the incisura.
E. The area between the midbrain and the free edges is divided into two spaces.

✅ **Answer E**
- The area between the midbrain and the free edges is divided into (a) an anterior incisural space located in front of the brainstem; (b) paired middle incisural spaces situated lateral to the midbrain; and (c) a posterior incisural space located behind the midbrain.

❓ **23. The tentorial incisura spaces, the FALSE answer is:**
- A. The anterior space extends upward around the optic chiasm.
- B. The middle space contains the interpeduncular and the chiasmatic cistern.
- C. The chiasmatic cistern communicates with the cisterna laminae terminalis.
- D. The middle space is related to the temporal part of the choroidal fissure.
- E. The posterior space is the site of the crural and ambient cisterns.

✅ **Answer B**
- The anterior incisural space contains the interpeduncular cistern, which is situated between the cerebral peduncles, and the chiasmatic cistern, which is located below the optic chiasm.

❓ **24. Basal cisterns, the FALSE answer is:**
- A. The ambient cistern is demarcated medially by the pulvinar.
- B. The ambient cistern is demarcated laterally by fimbria.
- C. The cisternal side of the choroidal fissure is located in the ambient cistern.
- D. The crural cistern cannot be reached through the choroidal fissure.
- E. The crural cistern can be exposed from the temporal horn.

✅ **Answer A**
- The ambient cistern is a narrow communicating channel demarcated medially by the midbrain, above by the pulvinar, and laterally by the parahippocampal and dentate gyri, and the fimbria of the fornix.

❓ **25. Quadrigeminal cistern, the FALSE answer is:**
- A. It is situated at the posterior incisural space.
- B. Its anterior and lateral walls junction lies at the choroid fissure.
- C. Its lateral walls separate the cistern from the occipital horn.
- D. Each lateral wall has anterior and posterior parts.
- E. The anterior lateral wall formed by the crus of the fornix.

✅ **Answer C**

- The lateral walls of the quadrigeminal cistern separate the cistern from the atria. Each lateral wall has anterior and posterior parts: the anterior part is formed by the crus of the fornix and the posterior part is formed by the part of the medial surface of the occipital lobe situated below the splenium.

❓ **26. The Quadrigeminal cistern, the FALSE answer is:**
   A.  The suprapineal recess of the third ventricle bulges into it.
   B.  Its lateral part of the anterior wall is formed by the part of the pulvinar.
   C.  Its medial part of the anterior wall is formed by the quadrigeminal plate.
   D.  Below colliculi, it extends into the cleft between the pons and cerebellum.
   E.  The trochlear nerve does not pass the quadrigeminal cistern.

✅ **Answer D**

- Below the colliculi, the cistern extends into the cleft between the midbrain and cerebellum called the cerebellomesencephalic fissure. This fissure cannot be reached through the choroidal fissure. The trochlear nerves arise below the inferior colliculi and course laterally around the midbrain and below the pulvinars to enter the ambient cistern.

❓ **27. The Quadrigeminal cistern, the FALSE answer is:**
   A.  Its roof is formed by quadrigeminal plate.
   B.  Its arachnoid envelope is continuous anteriorly with velum interpositum.
   C.  The venous structures are found in the superomedial part of the cistern.
   D.  The large arteries are found in the inferolateral part of the cistern.
   E.  It opens into ambient cistern and cavum vergae.

✅ **Answer A**

- The roof of the cistern is formed by the lower surface of the splenium and the broad membranous envelope that surrounds the great vein and its tributaries.

# Third Ventricle: Pathology—Tumors

*Baha'eddin A. Muhsen, Hazem Madi, Mustafa Ismail, Ignatius N. Esene, and Norberto Andaluz*


H. R. Salih et al. (eds.), *Cerebral Ventricles*, https://doi.org/10.1007/978-3-031-42342-0_6

**? 1. Third ventricle tumor epidemiology, the FALSE answer is:**
   A. Third ventricle tumors account for less than 1% of all brain tumors.
   B. A colloid cyst is the most common lesion in children.
   C. A colloid cyst occurs most commonly in the third ventricle of adults.
   D. A colloid cyst is most commonly located in the anterior part of the third ventricle.
   E. Ependymomas are the third most common primary third ventricular tumor found in children.

**✓ Answer B**
   – The most common intraventricular tumors in children are choroid plexus papillomas.

**? 2. Third ventricle tumor epidemiology, the FALSE answer is:**
   A. The most common true mass of the foramen of Monro in adults is a colloid cyst.
   B. The most common pediatric lesion of the foramen of Monro is the subependymal giant cell tumor.
   C. The subependymal giant cell tumor (WHO grade I) appears as a heterogeneously enhancing mass.
   D. The subependymal giant cell tumor is associated with Sturge Weber syndrome.
   E. The most frequent third ventricle floor lesion is the tuber cinereum hamartoma.

**✓ Answer D**
   – Subependymal giant cell tumor (WHO grade I) is associated with tuberous sclerosis.

**? 3. Third ventricle meningioma, the FALSE answer is:**
   A. Embryological origin can be from arachnoid cap cells.
   B. Embryological origin can be from ependymal lining cells.
   C. It typically has homogenous contrast enhancement on MRI.
   D. The most common location is the anterior third ventricle.
   E. Peak incidence is in the second decade.

**✓ Answer D**
   – The majority of third ventricular meningiomas are located posteriorly in the pineal region.

**4. Third ventricle meningioma, the FALSE answer:**
A. Commonly misdiagnosed as colloid cyst.
B. The majority are psammomatous meningioma.
C. Third ventricular meningiomas are more common in males.
D. In von Recklinghausen's disease, the incidence of intraventricular meningiomas is 16.6%.
E. Blood supply is from the medial posterior choroidal arteries.

**Answer B**
- The majority of third ventricle meningiomas are fibroblastic, syncytial, or of mixed type.

**5. Third ventricle colloid cyst, the FALSE answer is:**
A. Appear hyperdense on CT.
B. Are most commonly found near the foramen of Monro.
C. The cysts are lined with stratified squamous epithelial cells.
D. Can cause sudden death due to hypothalamic compression.
E. The most common presenting symptom is headache.

**Answer C**
- Colloid cyst is lined with an inner layer of cuboidal or cylindrical cells and an outer layer of connective tissue including vessels.

**6. Third ventricle colloid cyst, the FALSE answer is:**
A. It originates from the folding of the primitive neuroepithelium.
B. Shows homogenous enhancement on T1 MRI with gadolinium.
C. Contents of the cysts include old blood, foamy cells, fat, cholesterol crystals, and CSF.
D. Colloid cysts have varied appearances on MR images.
E. CT evidence of calcification in a colloid cyst is uncommon.

**Answer B**
- Colloid cyst shows peripheral enhancement on T1 MRI with gadolinium.

**7. Third ventricle arachnoid cyst, the FALSE answer is:**
A. Suprasellar arachnoid cysts arise from an imperforate membrane of Liliequist.
B. The wall of an arachnoid cyst is typically pale, translucent, and avascular.
C. Adult ventricular arachnoid cysts are very rare.

   D. Endoscopic cyst fenestration or excision are not treatment options.

   E. Third ventricular arachnoid cysts cause hydrocephalus.

**✓ Answer D**

— Endoscopic approaches might be considered owing to their being less invasive and giving a further chance of ETV.

**? 8. Third ventricle chordoid glioma, the FALSE answer is:**

   A. WHO grade II.

   B. Typical location is in the suprasellar region and the anterior part of the third ventricle.

   C. Homogenously enhancing on MRI.

   D. Vimentin is negative in chordoid glioma.

   E. GFAP is positive marker of chordoid glioma.

**✓ Answer D**

— GFAP and Vimentin are strongly positive immunohistochemical markers of Chordoid glioma in all reported cases.

**? 9. Third ventricle choroid plexus papilloma, the FALSE answer is:**

   A. Most commonly affect young children under the age of five.

   B. Tumor is neuroectodermal in origin.

   C. CPP commonly progresses to become malignant.

   D. High vascularity of the tumor makes GTR challenging.

   E. MRI appears as homogeneous or heterogeneous tumors with a cauliflower appearance.

**✓ Answer C**

— Choroid plexus papilloma is treated with surgery and rarely progresses to become malignant.

**? 10. Hamartomas of the tuber cinereum, the FALSE answer is:**

   A. Congenital, nonneoplastic heterotopias.

   B. Grossly may be pedunculated or sessile.

   C. Hypointense relative to gray matter on T1- and T2-weighted images.

   D. Associated with precocious puberty.

   E. Associated with gelastic epilepsy, resulting in spasmodic laughter.

✅ **Answer C**

- Hamartoma is isointense with gray matter on T1-weighted images (T1WI) and hyperintense relative to gray matter on the second echo of the T2-weighted images (T2WI) with no enhancing on MRI.

❓ **11. Pediatric third ventricle tumor, the FALSE answer is:**

A. Langerhans cell histiocytosis most commonly occurs in patients under the age of 2 years.
B. Germinoma peaks at around 10–12 years.
C. Germinoma arises at the anterior third ventricle more commonly than posterior aspect.
D. Hypothalamic-chiasmatic pilocytic astrocytoma has mild enhancement.
E. Craniopharyngioma peaks between the ages of 5 and 15 years.

✅ **Answer C**

- Germinoma arises at the anterior third ventricle less commonly than at the posterior aspect of the ventricle.

❓ **12. Adult third ventricle tumor, the FALSE answer is:**

A. Pituitary macroadenoma is the most common tumor affecting the anterior third ventricle.
B. Papillary craniopharyngioma rarely has calcifications.
C. Craniopharyngioma rarely manifests with hypopituitarism or diabetes insipidus.
D. Craniopharyngioma is more often a homogeneous solid mass.
E. Sellar meningiomas extending superiorly are distinguished by a dural base and sclerosis of skull base.

✅ **Answer C**

- Craniopharyngioma commonly manifests with hypopituitarism.

❓ **13. The posterior wall of the third ventricle tumors, the FALSE answer:**

A. Germinoma is the most common neoplasm.
B. Pediatric pineal tumors rarely cause hydrocephalus when the tumor is still small.
C. In children or adults, it can manifest as Parinaud syndrome.
D. Pineocytoma (WHO grade I) comprises mature cells mostly in teens and young adults.
E. Pineoblastoma (WHO grade IV) are highly malignant primitive tumors.

✓ **Answer B**
- Most pediatric pineal tumors manifest as third ventricle obstruction and hydrocephalus when the tumor is still small.

**? 14. Masses of the third ventricular floor, the FALSE answer is:**
A. Hamartoma of the tuber cinereum.
B. The most frequently seen lesion is lipoma of the third ventricle.
C. Basilar artery aneurysm.
D. Dermoid cyst.
E. Epidermoid.

✓ **Answer B**
- The most frequently seen lesion is a hamartoma of the tuber cinereum.

**? 15. The radiological appearance of third ventricle cyst, the FALSE answer is:**
A. Epidermoid cyst hypointense on T1WI, hyperintense T2WI with diffusion restriction.
B. Dermoid cyst hypointense on T1WI, hyperintense T2WI with diffusion restriction.
C. Colloid cyst hypointense on T1WI, hyperintense T2WI with homogenous enhancing.
D. Arachnoid cyst hypointense on T1WI, hyperintense T2WI without diffusion restriction.
E. Ependymoma hypointense (Solid part) on T1WI with restricted diffusion.

✓ **Answer C**
- Colloid cyst hyperintense on T1WI, hypointense T2WI with rim enhancing.

**? 16. Third ventricle ependymomas, the FALSE answer is:**
A. The majority has slow growth and remain asymptomatic till it is large enough at diagnosis.
B. Postoperative radiotherapy is the standard of care regardless of the extent of resection.
C. Anaplastic ependymoma has a high recurrence rate.
D. Ependymomas in the pediatric population are mostly supratentorial.
E. Ependymomas in the adult population are mostly supratentorial.

✅ **Answer D**

- At least half of ependymomas occur in the first two decades of life; they localize in the posterior fossa, particularly in small children. Ependymomas in the adult population are mostly supratentorial.

❓ **17. Third ventricle ependymomas, the FALSE answer is:**
   A. Common in neurofibromatosis type II.
   B. Ependymomas account for 2% of intracranial tumors in adults.
   C. Ependymomas account for 12% of intracranial tumors in children.
   D. Myxopapillary ependymoma is the most common histology.
   E. Subependymoma are extremely rare outside the ventricular system.

✅ **Answer D**

- Myxopapillary ependymoma is the most common histology in filum terminale.

❓ **18. Third ventricle central neurocytoma, the FALSE answer is:**
   A. Can be extra ventricular neurocytoma.
   B. WHO grade III.
   C. Synaptophysin is positive.
   D. Neuron-specific enolase is negative.
   E. MIB-1 LI is the most accurate to determine prognosis and tumor grade.

✅ **Answer B**

- Central neurocytoma is WHO II.

❓ **19. Third ventricle central neurocytoma, the FALSE answer is:**
   A. Attached to the septum pellucidum near the foramen of Monro.
   B. CT scans calcifications up to 50% of all cases.
   C. Neuronal nuclei are positive.
   D. Neurocytoma can occur extraventricularly.
   E. More common in females than males.

✅ **Answer E**

- Central neurocytomas are most prevalent among young adults. There is no specific correlation with genders.

**? 20. Third ventricle SEGA, the FALSE answer is:**
   A. Most commonly in the region of the foramen of Monro.
   B. The most common symptom is new onset or worsened seizures.
   C. SEGAs occur in 50% of tuberous sclerosis patients.
   D. The mean age of presentation is 13 years.
   E. Everolimus is used in the treatment of SEGA.

**✔ Answer C**
   - SEGAs occur in 5–15% of TSC patients. These tumors are important to recognize because of their strong association with TSC.

**? 21. Third ventricle subependymoma, the FALSE answer is:**
   A. WHO I.
   B. Its incidence in the third ventricle is higher than in the fourth ventricle.
   C. Most commonly occur in the fourth and fifth decades of life.
   D. Circumscribed non-enhancing mass lesions attached to the ventricular wall.
   E. Typically, patients are asymptomatic.

**✔ Answer B**
   - Subependymoma is most commonly seen in the fourth ventricle as follows: fourth ventricle: 50–60%, lateral ventricles (usually frontal horns): 30–40%, and third ventricle: rare.

**? 22. Third ventricle surgical approaches, the FALSE answer is:**
   A. Transcallosal is preferable in the absence of hydrocephalus.
   B. Transcortical approach has a risk of seizure of 5%.
   C. The supracerebellar infratentorial approach is best for colloid cysts.
   D. Occipital transtentorial approach used for posterior third ventricle tumors.
   E. The transcortical approach is favorable when hydrocephalus is present.

**✔ Answer C**
   - Parapineal lesions can be approached via a supracerebellar infratentorial approach or an occipital transtentorial approach.

**? 23. Approaches to the third ventricle, the FALSE answer is:**
   A. Transient mutism occurs as a result of bilateral cingulate gyrus retraction.
   B. Interforniceal approach can result in memory loss.
   C. In transcallosal approach, the opening is 2/3 anterior and 1/3 posterior to the coronal suture to avoid motor strip.
   D. Disconnection syndrome is more common with anterior callosotomy.
   E. Possible pitfall entering a cavum septum pellucidum in transcallosal approach.

**✓ Answer D**
- Disconnection syndrome is more common with posterior callosotomy (near the splenium) where more visual information crosses. The risk is reduced by creating a callosotomy <2.5 cm in length.

**? 24. Colloid cyst management, the FALSE answer is:**
   A. Bilateral VP shunt.
   B. Unilateral shunt with fenestration of the septum pellucidum.
   C. Surgical removal is recommended in patients <50 years at any cyst size with ventriculomegaly.
   D. A good result from stereotactic drainage is demonstrated by a hyperdense appearance on CT.
   E. Surgical removal is recommended when the cyst is >10 mm of size at any age.

**✓ Answer D**
- Unsuccessful stereotactic aspiration is demonstrated by hyperdensity on CT due to high viscosity. Low viscosity demonstrates a hypo- or isodense CT appearance.

# Third Ventricle: Pathology—Non-Tumors

Hernando Alvis-Miranda, Teeba A. Al-Ageely,
Aqeel K. Ibrahim, Salam N. Hussein,
Mostafa H. Algabri, Gulshan T. Muhammed,
and Mustafa Ismail

## Contents

7.1     Endoscopic Third Ventriculostomy – 76

**?  1.** **Congenital malformations, related to abnormal third ventricle contour, the FALSE answer is:**

A. Dilated third ventricle with congenital hydrocephalus is one infrequent cause.

B. Third ventricle atresia is rare, results from thalamic fusion.

C. Abnormal contour of the ventricle roof is seen with callosal dysgenesis.

D. Abnormal contour of the anterior aspect of the ventricle is seen with malformations of the neurohypophysis.

E. Congenital cysts may result in an abnormal ventricular contour or intraventricular masses.

**✓ Answer A**

— Dilation (due to hydrocephalus) is the most frequent abnormal contour pattern of the third ventricle.

**?  2.** **Massa Intermedia, the FALSE answer is:**

A. It is a commissure, due to reciprocal interhemispheric connections.

B. The MI comprises neurons and neuropil postulated to represent neuronal and glial progenitors.

C. Stria medullaris fibers that modulate motivation and mood may cross through the MI as suggested by DTI tractography studies.

D. MI size correlates with anterior thalamic radiation integrity.

E. MI size mediates the relationship between age and attention in healthy female subjects.

**✓ Answer A**

— It is not considered a commissure because no reciprocal interhemispheric connections have been found within it in humans. Nonetheless, its cytoarchitecture suggests that it is functionally active. The MI comprises neurons and neuropil with circularly oriented fibers postulated to represent neurosphere correlates (neuronal and glial progenitors).

**?  3.** **Massa intermedia volume loss and/or absence scenarios, the FALSE answer is:**

A. Congenital, decreased, or lack of the normal MI constraint leading to third ventriculomegaly.

B. Acquired, primary, or secondary MI volume loss associated with loss of cerebral volume.

   C. Acquired, increased third ventricular pressure/dilation causing stretching or rupture.
   D. Acquired, increased third ventricular pressure/dilation cannot cause rupture.
   E. Third ventriculomegaly can be a manifestation of a congenital malformation.

**✓ Answer D**

— It seems that MI volume loss and/or absence can occur in three main scenarios: (1) congenital, decreased or lack of the normal MI constraint leading to third ventriculomegaly; (2) acquired, primary or secondary MI volume loss associated with loss of cerebral volume; and (3) acquired, increased third ventricular pressure/dilation causing stretching or rupture

**? 4. Arachnoidal cysts, the FALSE answer is:**
   A. Arachnoid cysts represent 1% of intracranial lesions.
   B. They are mainly congenital.
   C. The wall of an arachnoid cyst is typically pale translucent and avascular.
   D. Most are symptomatic.
   E. Are thought to arise from an imperforate membrane of Liliequist.

**✓ Answer D**

— Most of the arachnoidal cysts are asymptomatic, however cognitive decline, hydrocephalus and focal neurological deficits have been reported.

**? 5. Arachnoidal cysts, the FALSE answer is:**
   A. The "ball valve" mechanism implies the secretion of fluid into the cyst.
   B. The clinical features may be due to obstructive hydrocephalus or intermittent foraminal obstruction
   C. Pendulous nature of the cyst can cause drop attacks or head bobbing.
   D. Endoscopic treatment offers the advantage of an additional third ventriculostomy.
   E. The size of endoscopic fenestrations is of little importance.

✅ **Answer E**

- MR-imaged CSF flow dynamics had demonstrated the importance of fenestrating suprasellar cysts endoscopically to the basal cisterns and not just to the ventricles to prevent secondary closure of the opening and recurrence of symptoms; wide fenestrations should be performed during endoscopic cystocisternostomies or ventriculocystostomies, with widening of the opening using a Fogarty catheter. The etiology of arachnoid cysts remains unclear. Various mechanisms have been proposed. These include a "ball valve" mechanism with only one-way flow of CSF into the cyst: secretion of fluid into the cyst or a slit-valve like structure of the arachnoid membrane that occurs around the basilar artery and opens and closes with arterial.

❓ **6. Cavernous malformations in and around the third ventricle, the FALSE answer is:**

    A. Symptoms are caused by mass effect from repetitive microhemorrhage and growth, sudden extracapsular hemorrhage, or seizures.
    B. For thalamic or basal ganglia lesions, rupture rates 2.8–4.1% per year.
    C. When symptomatic, without history of prior hemorrhage, are considered for microsurgical treatment.
    D. For thalamic or basal ganglia lesions re-rupture rates: 6.1–11% per year.
    E. Lesions unamenable to microsurgical resection are recommended for observation with serial imaging.

✅ **Answer C**

- Symptoms due to cavernous malformations in and around the third ventricle, are caused by mass effect from repetitive microhemorrhage and growth, sudden extracapsular hemorrhage with mass effect, or seizures. Thalamic or basal ganglia lesions tend to rupture between 2.8 and 4.1% per year, with re-rupture rates of 6.1–11% per year. Patients with symptomatic cavernous malformations, particularly with a history of prior hemorrhage, are considered for microsurgical treatment.

❓ **7. Cavernous malformations in and around the third ventricle, the FALSE answer is:**

    A. The anterior interhemispheric approach for medial thalamic lesions.
    B. The contralateral interhemispheric approach for lesions in the ventricle lateral wall.

   C. The ipsilateral interhemispheric approach for midline lesions in the basal thalamus.
   D. The lateral mid brain subtemporal approach for lesions with significant involvement of the pulvinar.
   E. The supracerebellar infratentorial approach for lesions in the basal thalamus.

✅ **Answer C**
- The ipsilateral interhemispheric approach is a better option to access midline lesions emanating into the third ventricle from the midline mesencephalon or hypothalamus.

❓ **8. Cerebral aqueductal stenosis, the FALSE answer is:**
   A. It manifests as dilatation of the third and lateral ventricles out of proportion to the fourth ventricle.
   B. Stenosis most often occurs in the distal aqueduct.
   C. May be congenital or acquired as a postinflammatory aqueductal gliosis.
   D. Aqueductal stenosis may manifest at any time from fetal age to adulthood.
   E. Up to 10% of adult hydrocephalus cases are due to aqueductal stenosis.

✅ **Answer B**
- Stenosis most often occurs in the proximal aqueduct.

❓ **9. Cerebral aqueductal stenosis, the FALSE answer is:**
   A. The aqueduct is the most common site of CSF intraventricular blockage.
   B. Responsible of 6–66% of cases of hydrocephalus in children.
   C. Responsible of 5–49% of cases of hydrocephalus in adults.
   D. There is only one peak of age distribution, in the first year of life.
   E. There is a bimodal peak of age distribution, first year and adolescence.

✅ **Answer D**
- There are two peaks of distribution for age, one in the first year of life and the other in adolescence.

❓ **10. Cerebral aqueductal stenosis, the FALSE answer is:**
   A. The sylvian aqueduct may become stenotic due to compression from mass lesions or intrinsic pathology.
   B. Intrinsic aqueductal stenosis may be congenital or acquired.

C. Non-tumoral aqueduct anomalies: stenosis, forking, septation, and gliosis.

D. Intraventricular hemorrhage (IVH) is the main etiology of aqueductal stenosis.

E. In most patients, the etiology is unknown.

✅ **Answer D**

— Three quarters of patients, the etiology of the disorder is not known (idiopathic aqueductal stenosis), In the remaining cases, it can be attributed to different causes: genetic factors (X-linked hydrocephalus), bacterial and viral infections (both intrauterine and infantile), hemorrhage (IVH of the prematurity, SAH), CNS malformations (Arnold-Chiari, spina bifida, encephaloceles, Dandy-Walker).

❓ **11. Cerebral aqueductal stenosis, the FALSE answer is:**

A. Aqueductal stenosis should not be considered a stable condition.

B. Partial stenosis is not influenced by functional mechanisms such as accumulation of fluid in the supratentorial ventricular system.

C. Aqueductal stenosis often is well tolerated for years.

D. Occlusion can be worsened by head traumas, SAH, or viral infections.

E. Aqueductal stenosis is a dynamic condition influenced by diverse CSF functional mechanisms.

✅ **Answer B**

— Partial stenosis may be completed by functional mechanisms such as accumulation of fluid in the supratentorial ventricular system that may cause distortion of the brain stem and kinking of the aqueduct, worsening the stenosis in a vicious cycle.

❓ **12. Cerebral aqueductal stenosis, the FALSE answer is:**

A. In normal circumstances, CSF flow in the aqueduct is turbulent.

B. CSF flow in the aqueduct is influenced by systole and diastole.

C. The shape of the aqueduct, driving a core of fluid centrally away from the wall effect, is important in maintaining a laminar flow.

D. Changes in aqueductal size and shape alter the volume flow rate and pattern.

E. As the aqueduct narrows, the flow decreases for a given pressure difference and the velocity increases.

✅ **Answer A**
- The movement of CSF through the aqueduct has a pulsatile nature, with a systolic and diastolic to-and-fro displacement.

❓ **13. Regard pressure gradient formation in aqueductal stenosis, the FALSE answer is:**
   A. May lead to typical third ventricle deformations.
   B. May lead to symptoms specific for hydrocephalus secondary to aqueductal stenosis.
   C. Are not related to gliosis formation and narrowing.
   D. May explain endocrinologic and visual disturbances due to mass effect on the chiasmatic hypothalamic region.
   E. May lead to global rostral midbrain dysfunction (mass effect on the quadrigeminal-pineal region).

✅ **Answer C**
- Some theories advocates that an increase of the pressure inside the aqueduct and of the wall shear stresses, gliosis with further narrowing may develop.

❓ **14. Cerebral aqueductal stenosis, the FALSE answer is:**
   A. When the aqueduct is stenotic, the systolic displacement of CSF is impaired.
   B. The CSF pulse during systole is transmitted to the ependymal walls.
   C. A pronounced CSF pulse during systole can lead to periventricular edema, ischemia, tissue damage, and ventricular dilatation.
   D. CSF flow through the aqueduct has a pulsatile nature, variations of diastole can lead to ventricular dilatation.
   E. CSF flow through the aqueduct has a pulsatile nature, variations of systole can lead to ventricular dilatation.

✅ **Answer D**
- The importance of CSF pulsation in the development of ventricular dilatation leads to the proposed pathophysiological mechanism: when the aqueduct is stenotic, the systolic displacement of CSF is impaired, thus, the pulse during systole is transmitted to the ependymal walls, leading to periventricular edema, ischemia, tissue damage, and ventricular dilatation.

**? 15. Cerebral aqueductal stenosis, the FALSE answer is:**
   A. In case of acute intraventricular block, the CSF cannot reach absorption sites.
   B. The ventricles increase in volume, but periventricular capillaries can absorb a significant part of CSF.
   C. Ventricular dilatation displaces the brain toward the skull and compresses the cortical veins.
   D. Development of venous congestion can further increase the ICP.
   E. Venous congestion and the brain swelling counteract the ventricular dilatation.

**✓ Answer B**
   — The ventricles increase in volume because the periventricular capillaries can only absorb part of the CSF produced.

**? 16. Regard to obstructive hydrocephalus natural history, the FALSE answer is:**
   A. Arteriolar/capillary regulated-CSF absorption in periventricular capillaries balances its production in isolated ventricles.
   B. If CSF pressure decreases an equilibrium cannot be established at near-normal pressure in the chronic phases.
   C. In chronic phase, the venous outflow compression and congestion disappears; with no force to counteract ventricular dilatation.
   D. As ventricles are isolated from subarachnoid spaces, the CSF pulse during systole increases transmantle pulsatile stress.
   E. Increased transmantle pulsatile stress continues to dilate the ventricles even if the mean CSF pressure is normal.

**✓ Answer B**
   — In the chronic phase of obstructive hydrocephalus, the CSF pressure decreases, creating an equilibrium established at near-normal pressure.

**? 17. ETV indications, the FALSE answer is:**
   A. Acquired aqueductal stenosis is a criterion.
   B. Adequate size of third ventricle (less 1 cm bicoronal diameter) is a criterion.
   C. Extension of the floor of the third ventricle behind and below the dorsum sellae is a criterion.
   D. A third ventricle >1 cm bicoronal diameter is a criterion.
   E. Potential patency of subarachnoid spaces is a criterion.

✅ **Answer B**
 — Adequate size of third ventricle (at least 1 cm bicoronal diameter) is a criterion for ETV.

❓ **18. Diagnosis of fetal aqueductal stenosis, the FALSE answer is:**
 A. ICP pressure results in compression of the cortical mantle, decreased vascular perfusion, and tissue ischemia.
 B. Dilated ventricles cause distention of the brain tissue, mechanical axonal shear, and gliosis.
 C. Prenatal diagnosis is often suspected at the time of the anatomic survey ultrasound (22–25 weeks of gestation).
 D. Ventriculomegaly from FAS has the appearance of overabundant CSF.
 E. Dilated ventricles and brain parenchyma compression are characteristics of overabundant CSF.

✅ **Answer C**
 — The prenatal diagnosis of FAS is often suspected at the time of the anatomic survey ultrasound at 18–22 weeks of gestation.

❓ **19. Diagnosis (typical findings on ultrasound) of fetal aqueductal stenosis, the FALSE answer is:**
 A. Asymmetric severe ventriculomegaly (greater than 15 mm).
 B. Dilated third ventricle (greater than 2 mm in transverse diameter).
 C. "Dangling" choroid plexus.
 D. Normal-appearing posterior fossa.
 E. Normal-appearing fourth ventricle.

✅ **Answer A**
 — Typical findings on ultrasound include bilateral symmetric severe ventriculomegaly (i.e., tricameral hydrocephalus), a dilated third ventricle, "dangling" choroid plexus, a normal-appearing posterior fossa and fourth ventricle, and normal to enlarged cranial biometry.

❓ **20. Patterns of the prenatal sonographic appearance of fetal aqueductal stenosis, the FALSE answer is:**
 A. Rapid progression in the third trimester.
 B. Progressive ventriculomegaly in the second and third trimesters.
 C. Slow progression in third trimester.
 D. Stable ventriculomegaly till end of pregnancy.
 E. Rapid progression till end of pregnancy.

✅ **Answer C**

- In a small sample size of eight subjects, Rault and colleagues described three patterns of the prenatal sonographic appearance of FAS: rapid progression in the third trimester, progressive ventriculomegaly in the second and third trimesters, and stable ventriculomegaly.

❓ 21. **Magnetic resonance imaging in fetal aqueductal stenosis, the FALSE answer is:**
   A. Axial, sagittal, and coronal T2-weighted single-shot fast spin-echo.
   B. True fast imaging with steady-state-free precession through the fetal brain.
   C. Diffusion-weighted imaging (DWI).
   D. Need to be a 3 T or greater MRI.
   E. T1-weighted spoiled gradient echo.

✅ **Answer D**

- Common fetal MRI protocols for ventriculomegaly include axial, sagittal, and coronal T2-weighted single-shot fast spin-echo and true fast imaging with steady-state-free precession through the fetal brain as well as DWI and T1-weighted spoiled gradient echo. Images were performed at a slice thickness between 3 mm with no skip in at least 1.5 T MRI.

## 7.1 Endoscopic Third Ventriculostomy

❓ 22. **ETV success definition, the FALSE answer is:**
   A. Improvement of clinical findings with no need for shunt placement.
   B. ICP despite ventriculomegaly with no need for shunt placement.
   C. ICP despite persistent short-term papilledema with no need for shunt placement.
   D. Persistence of bulging fontanel with need for shunt placement.
   E. Normal fontanel with no need for shunt placement.

✅ **Answer D**

- A successful ETV is defined as an improvement of clinical findings with no need for shunt placement.

**? 23. ETV failure, the FALSE answer is:**
    A. May be due to technical problems.
    B. Postoperative size of the ventricular system is a credible predictor for the successful procedure.
    C. Ventricular size is of secondary importance compared to clinical improvement.
    D. Predictive value of decrease in the ventricle size is unsatisfactory and unreliable.
    E. MRI flow studies are widely used and reliable objective postoperative tests for a successful ETV.
    F. May be due to inappropriate indication for the ETV.

**✓ Answer B**
    — Sometimes misleading, the postoperative size of the ventricular system is not always a credible predictor for the successful procedure. A blind study of 38 patients made by Buxton et al. showed that even though the third ventricular size was characteristically reduced after the ETV, the importance of these findings is of secondary importance compared to the clinical improvement. The predictive value of decrease in the ventricle size, especially during the early stage after the procedure, is unsatisfactory and unreliable.

**? 24. ETV outcomes, the FALSE answer is:**
    A. A 2-month-old male patient with post-infectious hydrocephalus and no history of shunt carries a 40% of successful outcome.
    B. A 5-year-old female patient with myelomeningocele and history of shunt carries a 50% of successful outcome.
    C. A 12-year-old male patient with pineal region germinoma and no history of shunt carries an 90% of successful outcome.
    D. A 9-year-old male patient with tectal glioma and no history of shunt carries an 80% of successful outcome.
    E. A 12-year-old female patient with aqueductal stenosis, and no history of shunt carries a 90% of successful outcome.

**✓ Answer A**
    — In 2010, an ETV success score (ETVSS) was introduced as a 6-month predictive model for the success of the procedure. Using only preoperative data, the score has been shown to predict the chances of the ETV success with close accuracy. In theory, the total of three scores

**◼ Table 7.1**  The endoscopic third-ventriculostomy success score

| Score | Age | Etiology | Previous shunt |
|---|---|---|---|
| 0 | <1 month | Post-infectious | Previous shunt |
| 10 | 1 m to <6 month | | No previous shunt |
| 20 | | Myelomeningocele; IVH; non-tectal brain tumor | |
| 30 | 6 m to <1 year | Aqueductal stenosis, tectal tumor, other etiology | |
| 40 | 1 year to <10 years | | |
| 50 | ≥10 years | | |

The ETVSS is calculated as age score + etiology score + previous shunt score

(age, etiology, shunt history) expressed as a percentage is the approximate chance of the ETV lasting 6 months without failure. Score of less than 40% relate to very low chance of success and a score of more than 80% correlate with a better chance of success compared to shunting. A 2-month-old male patient (10%) with post-infectious hydrocephalus (0%) and no history of shunt (10%) carries an 20% of successful outcome. In ◼ Table 7.1 is shown the Endoscopic Third-Ventriculostomy Success Score.

**? 25. ETV outcomes in children, the FALSE answer is:**
   A.  Late deterioration of the ETV occurs at a mean of 2.5 years after the ETV.
   B.  Second round failure at ~6 years after the procedure in children between 1 and 10 years old.
   C.  Common reasons for failure in early period: presence of a second membrane in the prepontine space, obliterated subarachnoid spaces.
   D.  Insufficient absorption of CSF is a cause of failure in early period.
   E.  The most frequent cause for late failure is the sealing of the fenestrated floor due to glial fibrosis.

✅ **Answer B**

– A population-based analysis showed the second-round failure at approximately 3 years after the procedure in children between 1 and 10 years of age with tumors or aqueductal stenosis.

❓ **26. Injuries of the third ventricle floor due to ETV, the FALSE answer is:**
    A. Damage to the pons.
    B. Damage to third cranial nerve.
    C. Serious tachycardia.
    D. Postoperative hyperkalemia.
    E. Damage to the hypothalamus.

✅ **Answer C**

– Injuries of the third ventricle floor might appear in the form of damage to the hypothalamus, pons, cerebral peduncle, and third cranial nerve, during the ETV, a serious bradycardia and postoperative hyperkalemia might occur due to distortion of the posterior hypothalamus. Other described complications include respiratory arrest, diabetes insipidus, epilepsy, pneumocephalus, subdural hematoma, and psychiatric disorders.

❓ **27. Negative prognosis indicators for ETV, the FALSE answer is:**
    A. Scarred membranes.
    B. Older age.
    C. Floppy premammillary membrane.
    D. Shunt history.
    E. Infection history.

✅ **Answer B**

– Apart from abnormal anatomy, other negative prognosis indicators might be existing scarred membranes, floppy premammillary membrane, young age, shunt history, infection, IVH, and myelomeningocele.

❓ **28. Basilar artery as landmark or as prognostic factor in ETV in children, the FALSE answer is:**
    A. Correct localization is a key prognostic factor.
    B. Missing an anteriorly placed basilar artery may have catastrophic consequences.
    C. ~72% of cases have a clear intraoperative visualization of the basilar artery.

D. Patients with a non-visible basilar artery the success rate exceeds 80%.

E. The basilar anatomy is a key prognostic factor.

**Answer D**

- One of the most important prediction factors for ETV in children is the correct localization of the basilar artery and its anatomy; missing an anteriorly placed basilar artery may have catastrophic consequences. Kulkarni et al. reported that among 309 successful cases, the rate of those with a clear intraoperative visualization of the basilar artery was 71.7% and among patients where basilar artery was not visible, the success rate was 47.4%.

**29. Influence of anatomical variations of the third ventricle in ETV, the FALSE answer is:**

A. A thin floor gives good visibility of major vessels beneath the floor of the third ventricle.

B. Small prepontine interval carries a very high risk of vascular injury.

C. Thickened/opaque floor hinders visibility of vascular structures in the interpeduncular cistern.

D. Diminished prepontine interval is not a contraindication for ETV.

E. Ballooning/herniating floor renders blunt perforation impossible or inefficient.

**Answer B**

- Patients with a diminished or absent prepontine interval and noncommunicating hydrocephalus should be considered as candidates for ETV. The procedure is safe and technically possible when supplemented with preoperative trajectory planning and intraoperative stereotactic guidance. The lasting success of ETV for controlling hydrocephalus in patients with a diminished PPI appears comparable to rates obtained from previous reports in patients with noncommunicating hydrocephalus.

**30. Influence of anatomical variations of the third ventricle in ETV, the FALSE answer is:**

A. Small foramen of Monro allows scope-normal range of endoscopic movements.

B. A narrow third ventricle restricts endoscopic movements.

C. Thickening anterior to mammillary bodies can—risk of basilar injury.

D. Abnormal interthalamic adhesions can interfere with endoscope manipulation.

E. Nonenlarged foramen of Monro despite chronic hydrocephalus—risk of forniceal damage.

✅ **Answer A**

— Nonenlarged foramen of Monro despite long-standing hydrocephalus difficult entry of scope and limits range of movement also it increases the risk of damage to fornices.

❓ **31. Anatomical variations of the Liliequist membrane, the FALSE answer is:**

A. Type A: two leaflets, diencephalic (DL) and mesencephalic leaves that originate at the dorsum sellae.

B. Type B LM: one leaf anteriorly and two leaflets posteriorly, splitting into DL and ML.

C. LM is found in most individuals, mainly separating the interpeduncular, prepontine, and chiasmatic cisterns.

D. Type C: a single mesencephalic membrane.

E. LM can be absent in 15.4 and 42.9%.

✅ **Answer D**

— Based on the various descriptions, LM has been divided into three types. In type A, the LM is composed of two leaflets, the diencephalic leaf and mesencephalic leaf (ML) that originate at the dorsum sellae. In type B, the LM appears as one leaf anteriorly and two leaflets posteriorly, with the LM arising as single membrane and then splits into DL and ML. In type C, the LM appears as a single (diencephalic) membrane.

❓ **32. Hypothalamic hamartomas (HH), the FALSE answer is:**

A. Rare, non-neoplastic heterotopic tissue containing normal neurons and glia but in an abnormal distribution.

B. Arise from the floor of the third ventricle, tuber cinereum, or mammillary bodies.

C. Although resemble tumors in certain ways, they do not tend to have a neoplastic evolution.

D. There is a moderate female predominance.

E. They grow at the same rate as the tissue from which they are derived.

**Answer D**
- The exact incidence of HH is unknown, but it has been estimated to be from 1 in 50,000–100,000 to 1 in 1,000,000 with a moderate male predominance.

**33. HH, the FALSE answer is:**
A. Gelastic seizures in HH are frequently pharmaco-refractory.
B. Ketogenic diet or both corticosteroids and KD are considered effective for gelastic seizures.
C. Leuprolide acetate plays a key role in medical therapy.
D. Resection of pedunculated lesions causing precocious puberty effectively reverses hormonal irregularities.
E. Medical therapy, the primary treatment for most patients with HH-related central precocious puberty.

**Answer B**
- Although exposing surgically resected neurons to ketone bodies may decrease spontaneous firing, possible efficacy of KD, both corticosteroids and KD are still considered ineffective. Medical therapy, including long-acting GnRH analogs such as leuprolide acetate, is now the primary treatment for most patients with HH-related CPP.

**34. Seizures in HH, the FALSE answer is:**
A. If a patient has both gelastic and dacrystic seizures, the cause is likely a HH.
B. If dacrystic seizures are not accompanied by gelastic seizures, the primary lesion is usually located in the frontal cortex.
C. A combination of dacrystic seizures and gelastic seizures usually occurs in younger age group.
D. Isolated dacrystic seizures usually occur in older age group.
E. In isolated dacrystic seizures, primary lesion is usually located in the temporal cortex.

**Answer B**
- Blumberg et al. demonstrated that if a patient has both gelastic and dacrystic seizures, the cause is likely a hypothalamic hamartoma. If dacrystic seizures are not accompanied by gelastic seizures, the primary lesion is usually located in the temporal cortex. A combination of dacrystic and gelastic seizures usually occurs in younger age group while isolated dacrystic seizures usually occur in older age group.

**?** **35. Seizures in HH, the FALSE answer is:**
   A. Seizure semiology may also include autonomic phenomenon, automatisms, epigastric auras, *déjà vu, déjà vécu,* crying, and motor symptoms.
   B. Seizure onset occurs in infancy but may occur as early as the neonatal period or as late as adolescence.
   C. Autonomic manifestations may be explained by changes in the limbic system, the adrenergic system, and the hypothalamic–pituitary axis.
   D. As years go by, the laughing attacks are less frequently associated with alteration of consciousness.
   E. Early gelastic seizures are associated with little or no impairment of consciousness.

**✓** **Answer D**
   — The early GS in infancy is associated with little or no impairment of consciousness, recurring daily. However, as years go by, the laughing attacks are more frequently associated with alteration of consciousness.

**?** **36. Clinical correlation in HH, the FALSE answer is:**
   A. Often associated with precocious puberty, behavioral disorders, and progressive cognitive decline.
   B. Smaller HH are associated with gelastic seizures.
   C. HH-related epilepsy occurs with medium to large sessile HH that are widely attached to the mammillary bodies and tuber cinereum.
   D. There may also be a direct relationship between the size of the HH and the severity of the epileptic syndrome.
   E. Tend to be smaller when causing isolated central precocious puberty.

**✓** **Answer B**
   — Size, shape, and location of HH can affect the clinical presentation. HH causing gelastic seizures tend to be larger than those associated with isolated central precocious puberty.

**?** **37. Central precocious puberty in HH, the FALSE answer is:**
   A. Average age of presentation in males is 3.7 years and in females is 2.5 years.
   B. Pedunculated HH located below the third ventricle tends to generate seizures, and commonly CPP.

   C. HH exerts physical pressure or neurosecretory influences leading to premature initiation of pulsatile GnRH release.
   D. HH itself can generate pulsatile GnRH release independently.
   E. Idiopathic CPP usually starts at the age of 5 years.

**✓ Answer B**

   – While idiopathic CPP usually starts at the age of 5 years, the average age of presentation for CPP related to HH in males is 3.7 years and in females is 2.5 years. Patients with pedunculated HH located below the third ventricle rarely present with seizures. Rather, these lesions are commonly associated with CPP. The pathophysiology of CPP is not fully elucidated. One hypothesis suggested that HH interacts with adjacent normal hypothalamic tissue by physical pressure or neurosecretory influences leading to premature initiation of pulsatile GnRH release by the normal hypothalamus. Other hypothesis suggested that the HH itself can generate pulsatile GnRH release independently, which in turn can stimulate the anterior pituitary.

**? 38. HH, the FALSE answer is:**
   A. Type I: small sessile lesions, horizontal plane of attachment to the tuber cinereum.
   B. Type II: mainly intraventricular lesions, vertical plane of insertion.
   C. Type III: combination of types I and II, both a horizontal and a vertical insertion plane.
   D. Type IV: very large, broad attachment to both the mammillary bodies and the hypothalamus.
   E. Type IV: extends into the interpeduncular cistern.

**✓ Answer A**

– Several surgical classifications have been proposed to orientate the best surgical approach and to predict the functional and epileptic outcome after the surgery. The original classification distinguished the peduncular from the sessile HH, posteriorly refined in two subcategories each one (1a and 1b for peduncular and 2a and 2b for sessile hamartomas). Arita proposed a close classification that separates the para-hypothalamic (usually the peduncular ones) from the intrahypothalamic (usually the sessile ones) HH.

– Delalande's classification into four subtypes remains the most used. Type I are small peduncular lesions characterized by a horizontal plane of attachment to the tuber cinereum; type II are mainly intraventricular lesions having a vertical plane of insertion; type III lesions are a combination of types I and II having both a horizontal and a vertical insertion plane; type IV lesions are generally very large tumors with broad attachment to both the mammillary bodies and the hypothalamus and have an extension into the interpeduncular cistern.

**39. HH, the FALSE answer is:**
   A.  Identifying the margins of the hamartoma is often challenging, particularly if it is broadly attached to the hypothalamus and mammillary bodies.
   B.  Sessile HH seem more amenable to complete resection than pedunculated ones.
   C.  Pterional approach is a feasible option for the para-hypothalamic subtypes.
   D.  Pterional approach seems best for targeting pedunculated HH that cause CPP.
   E.  Pedunculated HH seem more amenable to complete resection.

**Answer B**
   — Pedunculated HH seem more amenable to complete resection than sessile ones.

**40. HH, the FALSE answer is:**
   A.  Type I HH involves horizontal attachment to the hypothalamus and is best suited for a pterional approach.
   B.  When intraventricular component, the pterional approach may be combined with an endoscopic technique.
   C.  The frontotemporal approach carries a minor risk of complications.
   D.  Even a partial resection has been shown to relieve the endocrinological disturbance.
   E.  Type III HH can be managed with transcranial plus endoscopical technique.

**Answer C**
   — In general terms, despite excellent outcomes regard seizure control, the frontotemporal approach had resulted in major complications. Nevertheless, a pterional approach seems best for targeting pedunculated HH that cause endocrinological disturbances.

**41. Surgical management of HH, the false answer is:**
A. Extent of resection is correlated to attachment to the mammillary bodies (neuropsychological risk) and to the tuber cinereum (endocrine risk).
B. Pterional approach is the shortest and most direct route to the suprasellar cistern.
C. Morbidity includes transient third nerve palsy, thalamo-capsular infarcts, postoperative central diabetes insipidus, and hyperphagia.
D. Surgical resection of HH had succeeded in increasing the seizure-free rate above 80%, with low complication rate.
E. Access to the third ventricle and the intraventricular components is limited by the narrowness of the surgical corridor in pterional approach.

**Answer D**
- Although technological advances have been used, including perioperative MRI, and succeeded in increasing the seizure-free rate above 80%, the complication rate remains high, generally greater than 30%.

**42. Transcallosal transeptal interfornical approach for resection of HH, the FALSE answer is:**
A. Offers a wide exposition, with reduced risk of cerebral infarction and oculomotor nerve palsy.
B. There is risk of injury to the mammillary bodies and pituitary stalk, the inherent risk is a fornical damage causing memory deficit.
C. Seizure frequency reduction: 60–70%, seizure freedom 50 and 55%.
D. Common complications are thalamic infarction and memory impairment.
E. Avoid manipulation of the vessels in the suprasellar and interpeduncular cistern.

**Answer C**
- In some series, seizure frequency reduction was achieved in more than 90% and seizure freedom in between 50 and 55% of them (63, 64). Common complications are thalamic infarction and memory impairment (permanent in <10% of patients); however, counterbalanced by both cognitive and behavioral improvements (88%, 60%, respectively).

**? 43. Radiosurgery for HH, the false answer is:**
   A. Had a delayed action time.
   B. Median marginal dose is 17 Gy (range 14–20 Gy).
   C. 60–66% of seizure control.
   D. No impact on behavior.
   E. Sleep patterns and Electroencephalography (EEG) can reach normalization.

**✓ Answer D**
- There is increasing literature evaluating radiosurgery, two level 2 prospective studies reported respectively 60% and 66% of seizure control following radiosurgery. Furthermore, a progressive positive effect on behavior during the six following months, co-occurring with sleep patterns and EEG normalization, was described.

# Third Ventricle: Surgery

*Waeel O. Hamouda, Maliya Delawan, Aktham O. Al-Khafaji, and Mayur Sharma*

**? 1. Approaches to the third ventricular lesion located anterior to Foramen of Monro (anterior zone). False answer is**
   A. Transforaminal (transcortical, transcallosal, or endoscopic routes).
   B. Anterior interhemispheric translamina terminalis.
   C. Subfrontal translamina terminalis.
   D. Pterional translamina terminalis.
   E. Posterior interhemispheric transcallosal.

**✓ Answer E**

**? 2. Approaches to the third ventricular lesion located in the middle zone (between Foramen of Monro and massa intermedia). The false answer is:**
   A. Anterior interhemispheric transcallosal inter (trans)forniceal.
   B. Anterior interhemispheric transcallosal para(sub)forniceal supra-choroidal.
   C. Anterior interhemispheric transcallosal para(sub)forniceal sub-choroidal.
   D. Infratentorial supracerebellar.
   E. Extended transforaminal.

**✓ Answer D**

**? 3. Approaches to the third ventricular lesion located in the posterior zone (posterior to massa intermedia). The false answer is:**
   A. Posterior interhemispheric transcallosal.
   B. Occipital transtentorial.
   C. Transforaminal (transcortical, transcallosal, or endoscopic routes).
   D. Median supracerebellar infratentorial.
   E. Paramedian supracerebellar infratentorial.

**✓ Answer C**

**? 4. Approaches to lesions arising from the sella with suprasellar extension to the third ventricle. The false answer is:**
   A. Transsphenoidal (transnasal–transfrontal).
   B. Posterior interhemispheric routes.
   C. Pterional (optico-carotid, retro-carotid, prechiasmatic and lamina terminalis routes).

    D. Combined anterior interhemispheric/subfrontal/pterional with transcallosal.

    E. Subfrontal (through prechiasmatic and lamina terminalis).

✅ **Answer B**

— Anterior (not posterior) interhemispheric route can approach sellar/suprasellar lesions through the transcallosal, translamina terminalis, and prechiasmatic routes.

❓ 5. **Preoperative planning**

    **General considerations. The false answer is:**

    A. Prophylactic anticonvulsants are not mandatory in patients with a history of seizures.

    B. A Foley catheter is inserted to monitor for diabetes insipidus secondary to hypothalamus manipulation.

    C. Steroids.

    D. Prophylactic antibiotics.

    E. Pneumatic stockings for DVT prophylaxis.

✅ **Answer A**

— Although most approaches follow natural arachnoid spaces (e.g., subfrontal, trans-sylvian, interhemispheric, or supracerebellar), postoperative seizures still develop in 30% of cases due to retraction of the cerebral cortex or direct corticectomy as in the transcortical transforaminal approach.

❓ 6. **Preoperative planning**

    **Venous anatomy. The false answer is:**

    A. Preoperative magnetic resonance venography is essential.

    B. Magnetic resonance venography evaluates dural sinuses, bridging veins, and deep veins.

    C. Usually, the side with the smaller and fewer bridging veins should be retracted.

    D. Bridging veins are numerous at the anterior and posterior parts of the SSS.

    E. Veins dissection at their entrance into the cortex is easier than at their drainage into SSS.

✅ **Answer D**

— Significant bridging veins are much more frequently encountered in the middle part of SSS, usually 2–8 cm behind the coronal suture.

These veins should not be sacrificed for fear of disastrous cortical venous infarctions.

- However, most anterior and posterior parts of the sinus usually lack significant bridging veins.
- Preoperative magnetic resonance venography is essential to evaluate the size and, site of the dural sinuses (for burr holes placement), bridging veins (for choosing the side and, site of interhemispheric approaches), and deep veins (for choosing the most appropriate approach in occipital transtentorial or supracerebellar infratentorial approaches).

**7. Preoperative planning**

**CSF drainage for brain relaxation for third ventricle lesions. The false answer is:**

A. An EVD may be placed intraoperatively at the time of craniotomy.

B. Large cisterns are not accessible early in transcortical and interhemispheric approaches.

C. A lumbar catheter can be inserted before positioning in every surgery.

D. EVD can be left for monitoring hydrocephalus in the immediate postoperative course.

E. Ventricular tapping can be done from nearby frontal or occipital horns.

**Answer C**

- Lumbar drain should be avoided in third ventricular lesion cases with isolated supratentorial obstructed hydrocephalus to avoid transtentorial downward herniation.

**8. Operative planning**

**At the end of surgery. The false answer is:**

A. Remove cottonoids previously placed at the trigone or aqueduct opening.

B. Irrigating the ventricles with body-temperature Ringer lactate solution.

C. Removal of blood or trapped air prevents hydrocephalus or pneumocephalus.

D. Opening the septum pellucidum is not recommended.

E. Layer-by-layer inspection, from deep ependyma to superficial pia, ensures hemostasis.

✅ **Answer D**

- The septum pellucidum might be electively opened for a 5–8 mm hole (if not opened yet) to communicate with the contralateral lateral ventricle to prevent univentricular hydrocephalus and clean any blood clots or tumor debris in the contralateral lateral ventricle.

❓ **9. Operative planning**

**Tumor excision nuances. The false answer is:**

A. Usually, the surgical corridor is significantly smaller than the tumor.
B. The lesion must be removed in pieces and delivered into the field of view.
C. Central enucleation followed by peripheral dissection is performed.
D. Tumor vascular supply from choroid plexus is not exposed early in the dissection.
E. Simple cysts aspiration alone is an efficient treatment for intraventricular cystic lesions.

✅ **Answer E**

- Cysts walls have to be safely excised, if possible, to prevent a recurrence.

❓ **10. Transforaminal approaches**

**General considerations. The false answer is:**

A. Provide access to the anterior third ventricle.
B. Ipsilateral choroid plexus and thalamostriate vein run posteriorly toward the Foramen of Monro.
C. Used for tumors in the third ventricle extending to lateral ventricles through a wide Foramen of Monro.
D. Can be reached through transcallosal, transcortical, or endoscopic routes.
E. Need the foramen of Monro to be enlarged either by tumor or by hydrocephalus.

✅ **Answer B**

- The ipsilateral choroid plexus (medially) and the ipsilateral thalamostriate vein (laterally) run within the choroid fissure anteriorly toward the foramen of Monro where the thalamostriate vein turns medially to join the internal cerebral vein.

**? 11. Transforaminal approach through the Foramen of Monro**
**Surgical technique. The false answer is:**
A. From either side depending on surgeon handedness and hemispheric dominance.
B. Surgical tools with large diameters should not be introduced through the foramen.
C. Foramen of Monro can be further enlarged posteriorly by opening the choroidal fissure.
D. Foramen of Monro can be further safely enlarged anteriorly with transection of the forniceal column.
E. It may be difficult to visualize the origin of the colloid cyst from the roof of the third ventricle.

**✓ Answer D**
- Injury of the forniceal columns can lead to memory impairments.
- Alternatively, Foramen of Monro can be further enlarged posteriorly by opening the choroidal fissure as far posteriorly as the junction of the anterior septal vein with the internal cerebral vein located 3–13 mm posterior to the foramen of Monro in 40% of cases.

**? 12. Transcortical (trans-sulcal) transforaminal approach**
**General considerations. The false answer is:**
A. The transcortical approach utilizes the middle frontal gyrus.
B. Trans-sulcal approach utilizes the superior frontal sulcus.
C. It is easy to identify the superior frontal gyrus and sulcus.
D. Usually performed on the right side to avoid left dominant functional areas.
E. Left-handed patients need preoperative functional imaging to assess hemispheric dominance.

**✓ Answer C**
- The superior frontal gyrus can be very narrow, and the superior frontal sulcus can be interrupted so that clear identification may not be possible in every case.

**? 13. Transcortical (trans-sulcal) transforaminal approach**
**Surgical technique. The false answer is:**
A. Position supine.
B. Head elevated and turned 15° to the contralateral side.
C. Skin incision: modified short bi-coronal incision or a horseshoe incision.

    D. Craniotomy: quadrangular flap with the posterior edge at coronal suture and SSS exposed medially.

    E. Dura opened in a horseshoe fashion, based medially toward the SSS.

✅ **Answer D**

— There is no need to expose the SSS in the transcortical route to reduce the risk of sinus or bridging veins bleeding and injury.

❓ **14. Transcortical (trans-sulcal) transforaminal approach**
    **Specific complications. The false answer is:**

    A. A high incidence (10–30%) of postoperative seizures is due to traversing cortical tissues.

    B. The approach trajectory is essentially through a silent brain area.

    C. The false passage can be avoided by using neuronavigation.

    D. The false passage is avoided by tapping and localizing the anterior horn first with a brain cannula.

    E. Transient hemiparesis occurs with excessive posterior retraction on the premotor area.

✅ **Answer B**

— The trajectory disrupts commissural, projection, and short/long association fibers, which might lead to neurological deficits, e.g., mutism and confusion.

❓ **15. Anterior interhemispheric transcallosal approach**
    **Routes to the third ventricle. The false answer is:**

    A. Transthalamic.

    B. Transforaminal.

    C. Trans(inter)forniceal.

    D. Para(sub)forniceal suprachoroidal.

    E. Para(sub)forniceal subchoroidal.

✅ **Answer A**

❓ **16. Anterior interhemispheric transcallosal approach**
    **General considerations, the FALSE answer is:**

    A. The larger or the non-dominant (if both equal) lateral ventricle ought to be entered.

    B. Contraindicated in crossed dominance.

    C. The lateral position provides an interhemispheric route that needs minimal or no retraction.

D. The lateral position allows a side-by-side surgeon's hand position.
E. The lateral position has no drawbacks.

✓ **Answer E**

- In the lateral (park bench) position, there is a greater amount of midline distortion caused by gravity (midline subfalcine structures displaced inferiorly toward the ground), which reduces intraoperative orientation.

**17. Anterior interhemispheric transcallosal approach**
**Surgical technique (extradural part). The false answer is:**
A. Position supine or lateral park bench.
B. Head in a neutral position and flex if the patient is positioned supine.
C. Skin incision: modified short bi-coronal incision or horseshoe incision.
D. Craniotomy: quadrangular flap (4 cm behind and, 2 cm in front of the coronal suture).
E. Two burr holes were placed a few millimeters to the contralateral side of the midline exposing the SSS.

✓ **Answer D**

- Craniotomy main exposure is anterior to the coronal suture, i.e., extending 4 cm in front and 2 cm behind the coronal suture, as it is better not to extend the craniotomy far behind the coronal suture to avoid injuring the motor/the supplementary motor areas or the bridging veins during retraction.

**18. Anterior interhemispheric transcallosal approach**
**Surgical technique (interhemispheric part). The false answer is:**
A. Soft cotton balls replace hard self-retaining retractors.
B. Release of CSF from sulcal and callosal cisterns enhances brain relaxation.
C. The cingulate gyri are separated in the midline between the two callosomarginal arteries.
D. The deep edge of the falx at the level of the coronal suture marks the level of the upper surface of the corpus callosum.
E. The corpus callosum is incised in between the two pericallosal arteries.

✅ **Answer D**

- The deep edge of the falx at the level of the coronal suture usually terminates at the level of the cingulate gyrus or higher and does not reach the corpus callosum. Failure to identify and separate the cingulate gyri can lead to unneeded corticectomy of the cingulate gyri with the false surgical passage.
- The corpus callosum is differentiated from the kissing cingulate gyri by being deeper than the level of the inferior falcine edge, glistening white in color with a relatively hypovascular surface devoid of pial vessels.
- Soft cotton balls can be placed at the anterior and posterior extent of the interhemispheric dissection to maintain the working corridor without using hard self-retaining retractors.

❓ **19. Anterior interhemispheric transcallosal approach**
**Surgical technique (intraventricular part). The false answer is:**
A. Once subfalcine dissection and callosotomy have been done, a medial retractor can be used.
B. The ipsilateral ventricle is entered.
C. Thalamostriate vein located medial to choroid plexus.
D. For orientation, the choroid plexus is followed anteriorly till the foramen of Monro.
E. Incision of the septum pellucidum enhances CSF drainage.

✅ **Answer C**

- The thalamostriate vein is located lateral to the choroid plexus, i.e., on the right side of the plexus in the right lateral ventricle and on the left side of the plexus in the left lateral ventricle.

❓ **20. Anterior interhemispheric transcallosal approach**
**Specific complications, the FALSE answer is:**
A. Seen in up to 75% of patients, but most tend to resolve within 3 weeks.
B. Permanent cognitive or focal neurologic deficits are reported in 10% of cases.
C. The severity of disconnection symptoms correlates with the length of the callosotomy.
D. Contralateral hemiparesis or seizures can occur from retraction on the cerebral cortex.
E. Mutism is not a complication of the transcallosal approach.

✅ **Answer E**

- Acute mutism, which can range from mild slowness of speech initiation to frank complete mutism, can develop hours to days after surgery and can persist for several months—usually associated with lower extremity paresis, incontinence, disinhibition, and seizures.
- It can result from callosotomy or also from direct retraction of the anterior cingulate gyrus, septum pellucidum, and fornix or the supplementary motor area.
- Mutism is usually associated with anterior disconnection syndrome, "Split-brain syndrome" (i.e., left hemi-apraxia, left agraphia, left alien-hand syndrome).

❓ **21. Anterior transcallosal inter (trans) forniceal approach**
**General considerations. The false answer is:**
  A. Allow adequate exposure for most of the anterior and posterior parts of the third ventricle.
  B. Best used for patients who have a cavum septum pellucidum.
  C. Carries risk of potential bilateral injury of the fornices with memory impairment.
  D. Memory impairment usually affects recent memory and is usually transient.
  E. Midline forniceal dissection can extend up to 4 cm posterior to the foramen of Monro.

✅ **Answer E**

- Dissection should not exceed 2 cm posterior to the foramen of Monro to avoid injury to the hippocampal commissure, the injury of which may lead to significant memory impairment.

❓ **22. Anterior transcallosal inter (trans) forniceal approach**
**Surgical technique. The false answer is:**
  A. Midline forniceal raphe is identified by the site of septum attachment.
  B. If the septal leaves are fused, they must be sharply dissected to create a midline plane.
  C. The fornices are split in the midline along the interforniceal raphe.
  D. Underneath the fornices, the two veins of Rosenthal lie in the midline.
  E. Superior and inferior layers of tela choroidea are dissected to enter third ventricle roof.

✅ **Answer D**
- Underneath the fornices, within the velum interpositum, the two internal cerebral veins are usually widely separated, running posteriorly to join the vein of Galen.

❓ 23. **Anterior transcallosal para(sub)forniceal trans(supra)choroidal approach**
**General considerations. The false answer is:**
A. Suitable for gaining access to the anterior two-third of the third ventricle.
B. It entails dissecting the choroid fissure between the fornices and the thalamus.
C. If exposure of the posterior part of the lesion is inadequate, converse into the interforniceal route.
D. The suprachoroidal approach is medial to the subchoroidal approach.
E. Subchoroidal is considered safer than the suprachoroidal approach.

✅ **Answer E**
- The subchoroidal approach involves a higher possibility for the need to sacrifice the thalamostriate vein to untether the choroid plexus with the potential consequences of hemiplegia, mutism, and drowsiness.
- Suprachoroidal approach differs from the subchoroidal one in that it involves an incision in the Taenia fornicis superomedial to the choroid plexus of the lateral ventricle rather than the Taenia Thalami (Taenia choroidea) inferolateral to the choroid plexus.

❓ 24. **Anterior transcallosal para(sub)forniceal trans(supra)choroidal approach**
**Surgical technique. The false answer is:**
A. Dissection starts behind, where the anterior septal vein joins the internal cerebral vein.
B. Dissection continues through the "Tenia choroidea."
C. Incision of the superior membrane of the tela choroidea exposes velum interpositum.
D. The ipsilateral choroid plexus and internal cerebral vein are mobilized laterally.
E. Ipsilateral fornices are mobilized medially.

✅ **Answer B**
 − In the suprachoroidal approach, dissection is through the "Tenia fornices," lateral to the fornices and, medial to the choroid plexus to open the choroidal fissure. Dissection through the "Tinea choroidea" lateral to the choroid plexus and medial to the thalamostriate vein is for the subchoroidal approach.

❓ 25. **Anterior transcallosal para(sub)forniceal trans(supra)choroidal approach**
**Specific complications. The false answer is:**
A. Drowsiness, mutism, and hemiparesis occur from basal ganglia and thalamic infarctions.
B. Eye movement abnormalities and contralateral hemihypesthesia occur from tectal and medial thalamic lacunar infarctions.
C. Bleeding from the posterior aspect of the ventricle is poorly controlled through this approach.
D. Bleeding occurs during dissection between adherent internal cerebral veins (ICVs).
E. Dissection of choroid fissure anteriorly till foramen of Monro might injure the anterior septal vein.

✅ **Answer D**
 − The two internal cerebral veins are usually widely separated within the vallum interpositum by the bulging roof of the third ventricle.

❓ 26. **Anterior transcallosal para(sub)forniceal trans(supra)choroidal approach**
**Specific complications. The false answer is:**
A. Basal ganglia and thalamic infarctions occur from thalamostriate or ICVs injuries.
B. Tectal and medial thalamic infarctions occur from injury to plexal segment of the medial posterior choroidal artery.
C. Better convert to interforniceal route to control bleeding from posterior third ventricle.
D. No bridging vessels connect the two internal cerebral veins across the midline.
E. Injury to the anterior septal vein causes no morbidity.

✅ **Answer E**

- Excessive extension of the choroid fissure dissection anteriorly till the foramen of Monro might injure the anterior septal vein as it joins the internal cerebral vein causing venous infarction for the genu of the corpus callosum and deep medullary frontal white matter.

❓ **27. Anterior interhemispheric translamina terminalis approach**
**General consideration. The false answer is:**
  A. Ventricle exposure is limited by anterior commissure and by optic chiasm.
  B. A single bridging vein in front of the coronal suture can be arguably sacrificed if needed.
  C. Opened paranasal frontal sinuses must be "cranialized."
  D. The roof of deep non-cranialized parts of the paranasal frontal sinus is covered by periosteum.
  E. It is better to avoid opening the paranasal air sinuses during the craniotomy.

✅ **Answer E**

- The lower edge of the craniotomy should be leveled with the anterior cranial fossa base to increase the upward angle of vision and minimize the need for frontal lobes retraction; this will inevitably incorporate the paranasal frontal sinuses in the craniotomy bone flap.
- "Cranialization" of opened paranasal frontal sinuses is done by removing their posterior walls and stripping off their mucosa.

❓ **28. Anterior interhemispheric translamina terminalis approach**
**Surgical technique (extradural part). The false answer is:**
  A. Position supine.
  B. Head elevated in a neutral position and flexed.
  C. The skin incision is a bi-coronal one.
  D. Craniotomy is a midline frontobasal one.
  E. The anterior portion of the SSS is ligated and divided with the insertion of falx to crista galli.

✅ **Answer B**

- The head should be extended to allow backward falling of the frontal lobes by gravity, thus minimizing the need for retraction.

**29. Anterior interhemispheric translamina terminalis approach**
**Surgical technique (intradural part). The false answer is:**
A. The olfactory nerves are inevitably sacrificed.
B. Frontal lobes are gently elevated and separated.
C. The floor of the anterior cranial fossa is followed posteriorly to the sphenoid limbus.
D. Above chiasm, lamina terminalis is exposed by mobilizing anterior cerebral arteries.
E. Lamina terminalis is opened strictly in the midline.

**Answer A**
– The olfactory nerves can be freed from their arachnoid sleeve to the orbitofrontal cortex to allow more elevation of the frontal lobes without their sacrificing.

**30. Anterior interhemispheric translamina terminalis approach**
**Specific complications. The false answer is:**
A. Visual memory and olfaction deficits occur from injury to the anterior commissure.
B. Bitemporal hemianopia can occur from injury to the chiasma.
C. Ischemic events can occur from injury to the anterior communicating artery complex.
D. Failure to seal and clean the opened paranasal frontal sinuses has no consequence.
E. Anosmia can occur from injury to the olfactory nerve.

**Answer D**
– Failure to seal the paranasal frontal sinuses may precipitate for CSF rhinorrhea, while failure to clean their mucosal lining can lead to the formation of an intracranial mucocele.

**31. Lateral subfrontal translamina terminalis approach**
**General considerations. The false answer is:**
A. Useful for midline sellar and anterior third ventricular lesions.
B. Subfrontal space is maximized by extradural drilling of the irregular anterior fossa floor.
C. Through this approach, the whole third ventricle cavity is clearly exposed.
D. Sellar components of the tumor can be debulked first through the prechiasmatic space.
E. It can be performed through a cosmetic eyebrow incision.

✅ **Answer C**
- The ipsilateral superolateral wall of the third ventricle escapes direct visualization in the lateral subfrontal approach due to the inclined angle of vision.

❓ **32. Lateral subfrontal translamina terminalis approach**
**Surgical technique. The false answer is:**
A. Position supine.
B. Head extended and rotated 30° to the contralateral side.
C. Craniotomy: mini pterional preserving at least a 5 mm bony edge above the orbital rim.
D. Subfrontal arachnoid space is dissected till the ipsilateral anterior clinoid.
E. Above chiasm, the lamina terminalis is exposed by mobilizing anterior cerebral arteries.

✅ **Answer C**
- It is recommended that the craniotomy includes the ipsilateral orbital rim to increase the upward angle of vision and reduce frontal lobe retraction.

❓ **33. Pterional translamina terminalis approach**
**General considerations. The false answer is:**
A. For sellar, suprasellar, and anterior third ventricle lesions extending to parasellar area.
B. Sellar tumor part is reached through retro-carotid, optico-carotid, prechiasmatic routes.
C. Limitations are poor visualization of ipsilateral infero-anterior third ventricular lesions.
D. Combined with transcallosal approach for sellar lesions extending high in third ventricle.
E. Limitations are poor visualization of contralateral optico-carotid and retro-carotid spaces.

✅ **Answer C**
- A blind spot for this approach is for high third ventricular tumor extensions, especially in the supero-posterior part.

❓ **34. Posterior interhemispheric transcallosal approach**
**Surgical technique (extradural part). The false answer is:**
A. Position semi-sitting.
B. Head flexed in the neutral position.

  C. The skin incision is a horseshoe skin flap based medially.

  D. Craniotomy: parasagittal flap with inferior edge below external occipital protuberance.

  E. Dura opened in horseshoe fashion based medially toward SSS.

✅ **Answer D**

— The lower craniotomy edge does not cross below the transverse sinus, so it is usually planned 2 cm above the external occipital protuberance, and with two burr holes placed a few millimeters to the contralateral side of the midline exposing the SSS.

❓ 35. **Posterior interhemispheric transcallosal approach**

**Surgical technique (intradural part). The false answer is:**

  A. Interhemispheric arachnoid space is dissected deep down to the splenium.

  B. CSF can be released from sulcal and callosal cisterns to attain more brain relaxation.

  C. The deep edge of the falx terminates at the level of cingulate gyrus or higher.

  D. Cingulate gyri are dissected apart in the midline, exposing the posterior part of the corpus callosum.

  E. The corpus callosum is incised in between the two pericallosal arteries.

✅ **Answer C**

— The deep edge of the falx usually separates the cingulate gyri on the two hemispheres to reach and terminate at the upper surface of the splenium at the posterior portion of the corpus callosum, so it can be used to mark the level of the splenium.

❓ 36. **Posterior transcallosal approach**

**Specific complications. The false answer is:**

  A. Rarely seizures occur from excessive cortex retraction or bridging veins interruption.

  B. Visual field defects can occur from occipital lobe retraction.

  C. Posterior disconnection syndrome is characterized by preserving the ability to name objects.

  D. Mutism.

  E. Memory loss can occur from hippocampal commissure injury.

### ✅ Answer C
- Posterior disconnection syndrome is characterized by interruption of the ability of the splenium to transfer sensory input perceived in the right hemisphere from crossed senses like the left visual field and left-hand touch or uncrossed senses like right olfaction to the speech centers responsible for naming in the dominant left hemisphere.
- This will lead to left visual anomia, left tactile anomia, right olfactory anomia, and left hemialexia.

### ❓ 37. Occipital Transtentorial approach
**General considerations. The false answer is:**
A. An approach from the left side is preferred.
B. Suited for posterior third ventricular lesions with supra- and infratentorial components.
C. Suited for tumors displacing deep veins caudally or posteriorly.
D. Suited for tumors extending to the pineal region, posterior thalamus, trigone of the lateral ventricle, or medial temporal lobe.
E. Splenium can be retracted upward instead of being incised to allow extra exposure.

### ✅ Answer A
- An approach from the right side is preferred because a right hemianopia, resulting from a left-sided approach, produces greater difficulty with reading as the blind spot moves with the patient down the line of text, hiding what the patient is trying to read next, making it difficult to recognize the end of long words or even locate the end of the line.

### ❓ 38. Occipital transtentorial approach
**Surgical technique (extradural part). The false answer is:**
A. Position: three-quarter prone, concorde prone, or semi-sitting with the head flexed.
B. The semi-sitting with head flexed position is the preferred position.
C. Skin incision: trapdoor flap-based inferiorly.
D. Craniotomy: exposing posterior SSS, torcular Herophili, and ipsilateral transverse sinus.
E. Dura opened as two triangular flaps (base toward transverse and sagittal sinuses) or a single square flap (base toward the SSS).

**✅ Answer B**

- The three-quarter prone position with the right shoulder down is preferred as it allows the ipsilateral right occipital lobe to fall away by gravity, while the falx holds the contralateral left occipital lobe in place, thereby providing a parafalcine interhemispheric route that needs minimal or no retraction.

**❓ 39. Occipital transtentorial approach.**
**Surgical technique (tentorial part). The false answer is:**
A. Occipital bridging veins are usually absent.
B. The tentorium is transected parallel to the straight sinus.
C. The abducent nerve is mobilized and protected at the tentorial incisural edge.
D. Retraction of transected tentorium edge laterally and posteriorly enhances visualization.
E. The deep edge of the falx can be transected parallel to the straight sinus for more visualization.

**✅ Answer C**

- The trochlear nerve.
- The tentorium is transected in a posterior to the anterior direction in a line starting just anterolateral to the junction of the torcula and the straight sinus and running approximately 1 cm off midline parallel to the straight sinus toward the tentorial incisural edge exposing the upper surface of the cerebellum.
- The deep free edge of the falx can be transected for about 1–2 cm parallel to the straight sinus starting approximately 1 cm above the insertion of the vein of Galen into the straight sinus (after coagulating or clipping the inferior sagittal sinus), allowing retraction of the falx medially toward the contralateral side to provide a wider exposure.

**❓ 40. Occipital transtentorial approach.**
**Surgical technique (quadrigeminal part). The false answer is:**
A. Thick opaque quadrigeminal arachnoid must be cautiously opened by sharp dissection.
B. Third ventricle cavity is entered by opening the tela choroidea at the suprapineal recess.
C. Resection of the tumor is performed from below the ICVs and Rosenthal's basal veins.

D. Due to complex venous anatomy, paravenous or intervenous corridors can be used for piecemeal tumor removal.
E. Any tumor adherent to the veins is subtotally removed.

✅ **Answer C**

– Resection of the tumor is performed in the midline between the ICVs, Rosenthal's basal veins, splenial (posterior pericallosal) veins, and internal occipital veins on both sides. Resection of the tumor is performed inferior to Galen's vein, the ICVs, and Rosenthal's basal veins in the supracerebellar infratentorial approach.

❓ **41. Occipital transtentorial approach.**
**Specific complications. The false answer is:**
A. Cortical contralateral hemianopia can occur from ipsilateral occipital lobe injury.
B. Seizures rarely occur from occipital lobe retraction.
C. The only cause of postoperative eye movement abnormalities is an injury to the trochlear nerve.
D. The tentorium is well vascularized, and its transection might be very bloody.
E. The negligence of re-suturing the transected tentorium has not caused any complications.

✅ **Answer C**

– Postoperative eye movement abnormalities more commonly occur from injury to the quadrigeminal plate located in the depth of the surgical field.

❓ **42. Infratentorial supracerebellar approach.**
**General considerations. The false answer is:**
A. Suited for infratentorial tumors in the posterior third ventricle.
B. The surgical corridor can be median or paramedian.
C. It may be difficult in the presence of a steep tentorium.
D. Not optimally suited if the tumor infiltrates laterally or superiorly above the tentorium.
E. Preoperative echocardiography should be done to rule out patent foramen ovale.

**Answer A**

- Inferiorly or posteriorly displaced deep veins at the quadrigeminal cistern carries the risk for venous injury and incomplete lesion exposure by this approach. It is better suited when the deep veins are displaced superiorly or anteriorly.

**43. Infratentorial supracerebellar approach.**
**Surgical technique (positioning). The false answer is:**
A. The sitting position is preferred as it offers good exposure with minimal retraction.
B. Patients with a large body habitus benefit more from the prone position.
C. Three-quarter prone or prone positions are preferred in patent foramen ovale patients.
D. Three-quarter prone or prone positions are preferred in pediatric patients.
E. The head in the sitting position must be flexed to align the tentorium in the horizontal position.

**Answer B**

- Patients with a large body habitus benefit from the sitting position because high intrathoracic pressures in the prone position can complicate exposure due to venous hypertension.

**44. Infratentorial supracerebellar approach.**
**Surgical technique (extradural part). The false answer is:**
A. Position can be either sitting, three-quarter prone, or prone.
B. Skin incision: midline starting 3 cm above inion to C3 spinous processes.
C. Craniotomy: suboccipital, including the edge of foramen magnum inferiorly.
D. Dura is opened in a V or Y pattern with its base superiorly on the transverse sinus.
E. The occipital sinus is ligated and divided along with the underlying falx.

**Answer C**

- The craniotomy does not necessarily include the edge of the foramen magnum inferiorly unless there is associated Chiari I malformation but should expose the transverse sinus superiorly.

**? 45. Infratentorial supracerebellar approach.**
**Surgical technique (intradural part). The false answer is:**
A. CSF can be drained from near cisterns (e.g., cisterna magna) for brain relaxation.
B. Cerebellar bridging veins draining into the tentorium can be divided near the tentorium.
C. Inferior retraction may be applied on the bulging central culmen part of the vermis.
D. Thick opaque quadrigeminal arachnoid must be cautiously opened by sharp dissection.
E. The trajectory can be midline or lateral paramedian.

**✓ Answer B**
— Cerebellar bridging veins draining superiorly into the tentorium should be divided close to the cerebellum to prevent the retraction of inaccessible bleeding sources back into the tentorium.

**? 46. Infratentorial supracerebellar approach.**
**Surgical technique (quadrigeminal part). The false answer is:**
A. Posterior cerebral arteries branches can be distinguished from SCA branches by running above the tentorium.
B. The precentral cerebellar vein can be sacrificed to access the pineal region.
C. Third ventricle is accessed inferior to the pineal body at the level of the posterior commissure.
D. The tumor is dissected from the SCA and, vein of Rosenthal superolaterally, from internal cerebral vein superomedially, and from the colliculi and origin of trochlear nerves inferiorly.
E. After resection, communication with the third ventricle is confirmed by direct visualization.

**✓ Answer C**
— Third ventricle is accessed superolateral to the pineal body through the tela choroidea and velum interpositum at the level of the habenula and the suprapineal recess.

**? 47. Infratentorial supracerebellar approach.**
**Specific complications. The false answer is:**
A. Sitting position increases the risk of air embolism, tension pneumocephalus, and hyperflexion spine injury.
B. Open veins may be missing during hemostasis due to gravity-dependent collapse.

C. Postoperative foot drop occurs from sciatic nerve compression during a sitting position.
D. Both superior cerebellar and precentral veins can be sacrificed.
E. Disorders of eye movement occur from injury to the quadrigeminal plate.

**✓ Answer D**

- The superior vermian vein draining the superior cerebellar cortex and the precentral vein may connect to the Galen's vein separately or merge together into the "Superior Cerebellar Vein" before joining the Galen's vein. Don't sacrifice the superior cerebellar vein, as it might cause cerebellar venous infarction.
- An open bleeding vein may be missed during hemostasis due to gravity-dependent collapse of the bridging veins in a sitting position, which may reopen up in the postoperative period when the patient is made supine, and his blood pressure increases on the reversal from anesthesia.

**48. Endoscopic approach to the third ventricle.**
**Indications. The false answer is:**
A. Achieve large solid tumor excision.
B. Restore CSF pathways, e.g., ETV.
C. Obtain tumor biopsy, e.g., in pineal tumors.
D. Obtain CSF samples, e.g., in suspected pineal germ cell tumors.
E. Place intraventricular catheters under vision.

**✓ Answer A**

- Endoscopy can be used to achieve excision of small or cystic tumors considering the limited available space for surgical manipulation, e.g., a colloid cyst.

**49. Endoscopic approach to the third ventricle**
**Tumor excision limitations. The false answer is:**
A. Lesions of large sizes.
B. Lesions of vascular nature.
C. Lesions with soft suckable consistency.
D. Lesion engulfed by the choroid plexus.
E. Lesion covered by large adherent veins.

**✓ Answer C**

- Hard, fibrous consistency.

**?** 50. **Endoscopic approach to the third ventricle.**
**Advantages of using microscopy over endoscopy. The false answer is:**
A. 3D visualization without the need to wear special goggles.
B. Wider surgical corridor.
C. More chance of complete resection, especially large tumors.
D. More ability to control bleeding.
E. Better inspection around the third ventricle and the lesion.

**✓ Answer E**
— The most important advantage of the endoscope, besides being minimally invasive, is the ability to look around the third ventricle and around the tumor using different angled lenses in comparison to the limited straight angle of vision provided by the microscope.

**?** 51. **ETV.**
**Good prognostic factors. The false answer is:**
A. Late-onset (adolescent or adult).
B. Nontumoral obstructive hydrocephalus.
C. Previous SAH or meningitis.
D. Wide interpeduncular and prepontine cisterns.
E. Large ventricles.

**✓ Answer C**
— Previous SAH or meningitis might cause subarachnoid space scarring and obstruction interfering with CSF absorption at the arachnoid granulation level.

**?** 52. **ETV.**
**Relative contraindications. The false answer is:**
A. Slit-like ventricles.
B. Prior whole-brain radiation.
C. Communicating hydrocephalus.
D. Children below 10 years.
E. Prior meningitis.

**✓ Answer D**
— Children below 2 years might not benefit from ETV due to underdeveloped arachnoid granulations.

**?** 53. **Endoscopic approach to the third ventricle.**
**Surgical technique (extraventricular part). The false answer is:**
A. Position supine.
B. Head: neutral position and extended.

C. The skin incision is a 2–3-cm vertical incision based over the Kocher's burr hole.
D. The ipsilateral lateral ventricle is cannulated using a sheath with a ventricular introducer.
E. After the stylet is withdrawn, the endoscope is inserted through the sheath into the lateral ventricle.

**Answer B**

- The head is slightly flexed and elevated approximately 30° to prevent excessive CSF loss.

**54. Endoscopic approach to the third ventricle.**
**Surgical technique (intraventricular part). False answer is.**
A. Structures of the lateral ventricle are identified before advancing the endoscope through Foramen of Monro.
B. Mammillary bodies are identified as anterior landmarks, while the infundibular recess is identified as a posterior landmark of the area for ventriculostomy.
C. Ventriculotomy is made with the endoscopic instrument, monopolar cautery, or laser probe.
D. Ventriculostomy can be further dilated using inflating a small Fogarty.
E. Fluctuations in the floor of the third ventricle indicate free CSF flow.

**Answer B**

- The infundibular recess is identified as the anterior landmark, while mammillary bodies are identified as the posterior landmark of the area for ventriculostomy.

**55. Endoscopic approach to the third ventricle.**
**Specific complications, the FALSE answer is:**
A. Injury to the basilar artery leads to hemorrhage or pseudoaneurysm formation.
B. Failure to identify and open the Lillequist membrane below the floor of the third ventricle.
C. Excessive manipulation of the endoscope while in the third ventricle can damage the forniceal columns surrounding the foramen of Monro.
D. Large ventriculostomy fenestration.
E. A misplaced fenestration causes injury to the hypothalamus.

✓ **Answer D**
- Large fenestrations are always desirable. Failure to dilate the fenestration to an adequate size using the Fogarty balloon may threaten the healing and closure of the fenestration.
- The same occurs with failure to detect the Lilliquist membrane.
- An unrecognized delayed failure of ETV can lead to herniation and death.

**?** 56. **Surgery for colloid cyst of the third ventricle.**
   **The false answer is:**
   A. The episodically symptomatic colloid cyst should be managed conservatively.
   B. Its stalk is attached to the third ventricle roof.
   C. The only definitive treatment is resection by microsurgical or endoscopic techniques.
   D. CSF diversion procedures alone will not address the symptoms related to its mass effect.
   E. The cyst is opened and decompressed to reduce fornical manipulation and injury.

✓ **Answer A**
- A sudden increase in ICP from rapid-onset hydrocephalus can lead to sudden death. Any symptomatic colloid cyst should be excised if possible.

**?** 57. **Surgery for chordoid glioma of the third ventricle.**
   **The false answer is:**
   A. Rare incidence.
   B. WHO grade II tumors, so if not totally excised, it recurs.
   C. More common in females in the fifth decade of life.
   D. Mostly located in the posterior part of the third ventricle.
   E. Present with obstructive hydrocephalus or mass effect.

✓ **Answer D**
- Mostly located in the anterior part of third ventricle in the midline.

**?** 58. **Approaches to third ventricle.**
   **Postoperative complications. The false answer is:**
   A. Postoperative permanent morbidities vary from 5 to 50%, depending on the approach.
   B. 30-day postoperative mortality at present is around 8%.

C. Early causes of mortality are usually vascular or uncontrolled high ICP.
D. Seizures, subdural collection, and hydrocephalus can be a potential complication.
E. Postoperative polyuria is not related to surgery.

✓ **Answer E**

- Pituitary/hypothalamic dysfunction might occur because of surgical manipulation or injury. Management includes steroid with or without desmopressin acetate (DDAVP) administration, close monitoring of urine output, urine specific gravity, and serum sodium.

? **59. Air embolism.**

**General considerations. The false answer is:**
A. More with sitting position.
B. A patent foramen ovale should be excluded preoperatively by echocardiography.
C. The fatal dose of air embolism is about 20 cc.
D. Presents with acute hypotension and arrhythmia.
E. Occur from air suction into negatively pressured venous channels.

✓ **Answer C**

- The dose causing a fatal clinical sequela similar to other pulmonary embolisms is about 200–400 cc depending on the rate of air accumulation.

? **60. Air embolism.**

**Preoperative preparations. The false answer is:**
A. Transesophageal or precordial echocardiography probe.
B. Large peripheral cannula.
C. Capnogram for end-tidal $CO_2$ monitoring,
D. PEEP.
E. Compression stockings.

✓ **Answer B**

- A central venous line with a multichannel catheter long enough to reach down to the right atrium is mandatory to allow aspiration of sizable air emboli from the atrium.

? **61. Air embolism.**

**Intraoperative management. The false answer is:**
A. The venous sinuses should be covered with wet laparotomy pads soaked with saline.

B. The field flooded with saline irrigation.

C. Attempts of air aspiration from the right atrium using the CVL.

D. Patient's head lowered (Trendelenburg position) if possible.

E. The preferred position is right decubitus.

✅ **Answer E**

— The preferred position (if possible) in case of air embolism is the left decubitus position, as it keeps the air high up in the right atrium away from the pulmonary arteries.

❓ **62. Approaches to third ventricle.**
**Postoperative seizures. The false answer is:**

A. More common with transcallosal than transcortical procedures.

B. Occur in one-third of cases undergoing transcortical approach.

C. It may be caused by retraction injuries to the brain during interhemispheric approaches.

D. Electrolyte imbalance from hypothalamic dysfunction may cause refractory seizures.

E. Confined intraventricular lesions uncommonly present with seizures.

✅ **Answer A**

— Seizures are less common with interhemispheric transcallosal approaches as they avoid traversing cortical tissues.

❓ **63. Approaches to third ventricle.**
**Postoperative subdural collection. The false answer is:**

A. Occur in approximately 40% of patients.

B. More than 50% of these subdural collections will eventually require surgical drainage.

C. Occur due to rapid collapse of the hemisphere from ventricular CSF drainage.

D. Subdural hematomas are more common than subdural hygromas.

E. Over-shunting adds to the problem of postoperative subdural collection.

✅ **Answer B**

— Only one-fourth of these subdural collections (10% of all operated patients) will require surgical drainage.

❓ **64. Approaches to third ventricle.**
**Postoperative subdural collection risk factors. The false answer is:**

A. More pronounced preoperative ventriculomegaly.

   B. Sitting position.
   C. Excessive use of mannitol.
   D. Delayed postoperative shunt placement.
   E. Elderly patients.

✅ **Answer D**

- Preoperative or intraoperative shunt placement is associated with an increased risk of developing subdural collections.
- Temporary preoperative closure of previously placed shunts (i.e., sigma loop tying of the distal catheter or setting a programmable shunt to very high pressure) might reduce the risk of postoperative subdural collections.

❓ **65. Approaches to third ventricle.**

**Postoperative hydrocephalus. The false answer is:**

   A. Preoperative hydrocephalus is common with third ventricular tumors.
   B. Resection of the third ventricular tumor prevents the persistence of preoperative hydrocephalus.
   C. 25% of shunted cases experience shunt obstruction.
   D. In approaches to the anterior third ventricle, the anterior burr hole should be drilled electively.
   E. In approaches to the posterior third ventricle, the posterior burr hole should be drilled electively.

✅ **Answer B**

- Unfortunately, one-third of operated patients continue to have persistent hydrocephalus postoperatively despite gross total tumor resection.
- In approaches to the anterior third ventricle (anterior interhemispheric or transcortical) or posterior third ventricle (supracerebellar infratentorial or occipital transtentorial or posterior interhemispheric) where an EVD was not placed at the end of the surgery, a rescue burr hole (Kocher or Dandy/Fraizer, respectively) should be drilled intraoperatively to have immediate access for CSF drainage if acute postoperative hydrocephalus occurs.

# Fourth Ventricle: Anatomy

*Hayder R. Salih, Maria L. Laffitte, Naba G. Husseini, Saleh A. Saleh, Mohammed S. Altaee, and Samer S. Hoz*



H. R. Salih et al. (eds.), *Cerebral Ventricles*, https://doi.org/10.1007/978-3-031-42342-0_9

**1. The fourth ventricle, the FALSE is:**
   A.  Tent-shaped midline cavity located between the cerebellum and the brainstem.
   B.  It is dorsal to pons and medial to cerebellar peduncles.
   C.  It is connected through the aqueduct with the third ventricle.
   D.  It is connected through foramen of Magendie with cisterna magna.
   E.  Most of the cranial nerves arise near its roof.

**Answer E**
   — Most of the cranial nerves arise near its floor. It has a roof, a floor, and two lateral recesses.

**2. The fourth ventricle, the FALSE is:**
   A. Roof expands laterally and posteriorly just below the aqueduct.
   B. At lateral recess level, it has the greatest height and width.
   C. Apex of its roof is called the fastigium.
   D. Inferior fastigium is thicker than superior part.
   E. Superior part of fastigium is formed by neural structures.

**Answer D**
   — The superior part is distinctly different from the inferior part, in that the inferior part is formed largely by thin membranous layers and the superior part is formed by thicker neural structures.

**3. The fourth ventricle roof, the FALSE is:**
   A. Its superior part is divided into two lateral and single median part.
   B. Median parts are formed by the superior medullary velum.
   C. Lateral parts are formed by inner surface of cerebellar peduncles.
   D. Superior medullary velum is continuous at fastigium with inferior medullary velum.
   E. Caudal portion of each lateral wall is formed by medial surface of superior cerebellar peduncle.

**Answer E**
   — The rostral portion of the ventricular surface of each lateral wall is formed by the medial surface of the superior cerebellar peduncle, and the caudal part is formed by the inferior cerebellar peduncle.

**4. Cerebellar peduncle and fourth ventricle, the FALSE is:**
   A. Middle peduncle is separated from ventricular surface by inferior and superior peduncles.

B. Fibers of inferior peduncle line ventricular surface of superior margin of lateral recess.
C. Fibers of inferior cerebellar peduncle ascend in the posterolateral midbrain.
D. Fibers of superior peduncle arise in dentate nucleus.
E. Fibers of superior peduncle form ventricular surface of superior part of lateral wall.

✅ **Answer C**
- The fibers of the inferior cerebellar peduncle ascend in the posterolateral medulla and turn posteriorly in the inferomedial part of the fiber bundle formed by the union of the three peduncles.

❓ 5. **Upper part of fourth ventricle roof, the FALSE is:**
A. Cisternal surface of the roof form the anterior wall of the cerebellopontine fissure.
B. Cerebellomesencephalic fissure is also called as precentral cerebellar fissure.
C. Lingula is a thin narrow tongue of vermis sits on surface of superior medullary velum.
D. Interpeduncular sulcus marks junction of superior and middle cerebellar peduncles.
E. Trochlear nerves arise in cerebellomesencephalic fissure below inferior colliculi.

✅ **Answer A**
- The cisternal (external) surface of the structures forming the superior part of the roof also form the anterior wall of the cerebellomesencephalic fissure.

❓ 6. **Inferior part of fourth ventricle roof, the FALSE is:**
A. It slopes sharply ventral and slightly caudal from fastigium.
B. Ventricular and cisternal surfaces are formed by same structures in caudal midline.
C. Ventricular surface is formed by nodule, inferior medullary velum, and tela choroidea.
D. Inferior medullary velum blends into dorsal margin of each lateral recess.
E. Inferior medullary velum is continuous at level of lateral recess with superior medullary velum.

✅ **Answer E**

- The inferior medullary velum is continuous at the level of the fastigium with the superior medullary velum.

❓ 7. **Tela choroidea of fourth ventricle, the FALSE is:**

A. It forms caudal part of inferior portion of roof and inferior wall of each lateral recess.
B. Consists of two thin, semitransparent layers, with thickness comparable to arachnoid.
C. Choroid plexus projects from ventricular surface of tela choroidea into fourth ventricle.
D. The telovelar junction extends from the nodule into each lateral recess.
E. It is completely enclosing inferior half of the fourth ventricle.

✅ **Answer E**

- The tela choroidea does not completely enclose the inferior half of the fourth ventricle but has three openings into the subarachnoid space: the paired foramina of Luschka located at the outer margin of the lateral recesses and the foramen of Magendie located at the caudal tip of the fourth ventricle.

❓ 8. **Cerebellomedullary fissure, the FALSE is:**

A. It faces and related to cisternal surface of caudal half of fourth ventricle roof.
B. Its dorsal wall is formed by uvula, tonsils, and biventral lobules.
C. Between tonsil, tela choroidea, and inferior medullary velum, it is called telovelotonsillar cleft.
D. It extends superiorly to level of lateral recesses and communicates with cisterna magna.
E. Extension of telovelotonsillar cleft over upper pole of tonsil is called peritonsillar cleft.

✅ **Answer E**

- The portion of the fissure between the tonsil, the tela choroidea, and the inferior medullary velum is called the telovelotonsillar cleft, and the superior extension of this cleft over the superior pole of the tonsil is called the supratonsillar cleft.

❓ 9. **Lateral recesses of fourth ventricle, the FALSE is:**

A. They are formed by union of fourth ventricle roof and floor.
B. They extend laterally above the cerebellar peduncles.

C. Ventral wall of each is formed by junctional part of floor and the rhomboid lip.
D. They open through the foramina of Luschka into the cerebello-pontine angles.
E. Rostral wall of each is formed by caudal margin of cerebellar peduncles.

✓ **Answer B**

— They extend laterally below the cerebellar peduncles and open through the foramina of Luschka into the cerebellopontine angles.

**? 10. The fourth ventricle floor, the FALSE is:**
A. The inferior cerebellar peduncle courses upward in the floor.
B. Peduncle of flocculus interconnecting inferior medullary velum and flocculus crosses.
C. Fibers of vestibulocochlear nerve arise ventral to lateral recess.
D. Rootlets of glossopharyngeal nerve arise ventral to lateral recess.
E. Rootlets of vagus nerve arises ventral to lateral recess.

✓ **Answer C**

— The fibers of the vestibulocochlear nerve cross the floor of the recess.

**? 11. The cerebellopontine fissure, the FALSE is:**
A. Lateral recess opens into cerebellopontine angle along cerebellopontine fissure.
B. It is formed by folding of vermis around lateral side of inferior cerebellar peduncle.
C. Middle cerebellar peduncle fills the interval between its two limbs.
D. Lateral recess and foramen of Luschka open into medial part of the inferior limb.
E. Trigeminal nerve arises from pons along the superior limb of the fissure.

✓ **Answer B**

— The cerebellopontine fissure is a V-shaped fissure formed by the folding of the cerebellar hemisphere around the lateral side of the pons and the middle cerebellar peduncle.

**? 12. Flocculus, the FALSE is:**
A. Vestibulocochlear nerves enter brainstem anterosuperior to the flocculus.
B. Facial nerves enter brainstem anterosuperior to the flocculus.

C. The fila of the glossopharyngeal nerve cross anteroinferior to the flocculus.
D. The fila of the vagus nerve enter the brainstem anterosuperior to the flocculus.
E. Flocculus projects into cerebellopontine angle at confluence of CPF and CMF.

**Answer D**

- The flocculus projects into the cerebellopontine angle at the confluence of the cerebellopontine and cerebellomedullary fissures. The vestibulocochlear and facial nerves enter the brainstem anterosuperior to the flocculus, and the fila of the glossopharyngeal and the vagal nerves cross anteroinferiorly to it.

**13. The choroid plexus of fourth ventricle, the FALSE is:**

A. Its entire shape presents the form of a letter T, vertical limb of which is double.
B. Its medial segments are located in floor near the midline.
C. Medial segment is divided into rostral or nodular part and caudal or tonsillar part.
D. Lateral segment extends parallel to telovelar junction into cerebellopontine angles.
E. Lateral segment is anterior to tonsils, extends inferiorly through foramen of Magendie.

**Answer E**

- The tonsillar parts of the medial segment are anterior to the tonsils and extend inferiorly through the foramen of Magendie.

**14. The fourth ventricle, the FALSE is:**

A. Rostral two-third of floor are posterior to pons and caudal one-third is posterior to medulla.
B. Cranial apex of the floor is at the level of the cerebral aqueduct.
C. The floor is divided longitudinally into symmetrical halves by the sulcus limitans.
D. The obex is located at rostral end of remnant of the spinal canal.
E. Line connecting orifices of lateral recesses is located at the junction of pons and medulla.

**Answer C**

- The floor is divided longitudinally from the rostral apex to the caudal tip into symmetrical halves by the median sulcus.

**? 15. The floor of the fourth ventricle, the FALSE is:**
    A. Superior part apex is at cerebral aqueduct.
    B. Intermediate part of extends into the lateral recesses.
    C. Inferior part has a triangular shape and is limited laterally by foramina of Luschka.
    D. It is divided longitudinally symmetrical halves by the median sulcus.
    E. Sulcus limitans divides each half of floor into median eminence and vestibular area.

**✓ Answer C**
    — The inferior part has a triangular shape and is limited laterally by the taeniae marking the inferolateral margins of the floor. Its caudal tip, the obex, is anterior to the foramen of Magendie.

**? 16. The fourth ventricle floor, the FALSE is:**
    A. Sulcus limitans is most prominent in pontine and medullary portions of floor.
    B. Sulcus limitans deepens at two points to form dimples called foveae.
    C. Facial colliculus overlies the abducent nucleus.
    D. Hypoglossal triangle is medial to superior fovea and overlies hypoglossal nucleus.
    E. The funiculus separans crosses the lower part of the vagal triangle.

**✓ Answer D**
    — The hypoglossal triangle is medial to the inferior fovea and overlies the nucleus of the hypoglossal nerve. Caudal to the inferior fovea and between the hypoglossal triangle and the lower part of the vestibular area is a triangular dark field, the vagal triangle, that overlies the dorsal nucleus of the vagus nerve.

**? 17. Posterior circulation and fourth ventricle, the FALSE is:**
    A. SCA is intimately related to the superior half of the roof.
    B. Posterior inferior cerebellar artery (PICA) is intimately related to the inferior half of the roof.
    C. PICA supplies choroid plexus in the floor.
    D. Anterior inferior cerebellar artery (AICA) is intimately related to the lateral recess and the foramen of Luschka.
    E. AICA supply portion of choroid plexus in cerebellopontine angle and adjacent part of lateral recess.

✅ **Answer C**
- The PICA supplies the choroid plexus in the roof and the medial part of the lateral recess.

❓ **18. The fourth ventricle and venous system, the FALSE is:**
  A. There are no major veins within the cavity of the fourth ventricle.
  B. Veins of superior cerebellar peduncle course on superior part of floor.
  C. Veins of inferior cerebellar peduncle drain the inferior half of the roof.
  D. Veins of the middle cerebellar peduncle drain lateral wall.
  E. Veins of middle cerebellar peduncle drain cerebellopontine angle around lateral recess.

✅ **Answer B**
- The veins most intimately related to the fourth ventricle are those in the fissures between the cerebellum and the brainstem and on the cerebellar peduncle. The veins of the cerebellomesencephalic fissure and the superior cerebellar peduncle course on the superior part of the roof, the veins of the cerebellomedullary fissure and the inferior cerebellar peduncle drain the inferior half of the roof, and the veins of the cerebellopontine fissure and the middle cerebellar peduncle drain the lateral wall and the cerebellopontine angle around the lateral recess.

# Fourth Ventricle: Pathology—Tumors

*Ali A. Dolachee, Abdullah K. Al-Qaraghuli,
Sajjad G. Al-Badri, Alkawthar M. Abdulsada,
Hagar A. Algburi, Zinah A. Al-araji,
and Mustafa Ismail*

**? 1. MB, the FALSE answer is:**
   A. Most common malignant pediatric brain tumor.
   B. Accounts for 10–20% of brain tumors in children.
   C. Observed frequently in adults.
   D. Exhibits bimodal distribution in children, peaks at 3–4 years and at 8–9 years.
   E. Male gender predilection with a male-to-female ratio of 1.5:1.

**✓ Answer C**
   − Although MB is the most common malignant neoplasm of the CNS in children, it is a rare disease in adults.

**? 2. MB gross pathology, the FALSE answer is:**
   A. Pinkish gray to purple mass.
   B. Commonly arising from the medullary velum.
   C. Some lesions are firm discrete masses, others are soft and friable.
   D. In a majority of cases, tumor invades floor of the fourth ventricle.
   E. A white, "sugar-coated" appearance of the cerebellar surface.

**✓ Answer D**
   − In a minority of cases, the tumor invades the floor of the fourth ventricle and its blood supply from the PICA.

**? 3. Microscopically, classic MB appearance, the FALSE answer is:**
   A. A dense sheet of small, basophilic cells with little cytoplasm.
   B. Round to oval hyperchromatic nuclei.
   C. A high mitotic index may be seen.
   D. Evidence of neuronal or glial differentiation is noted in up to 50% of cases.
   E. Homer Wright rosettes are infrequently present.

**✓ Answer E**
   − Homer Wright rosettes are commonly present.

**? 4. Histologic subtypes of MB, the FALSE answer is:**
   A. These are classic, anaplastic, large cell, desmoplastic/nodular, and extensive nodularity.
   B. All are (WHO) grade IV tumors.
   C. Anaplastic MB is characterized by nuclear pleomorphism, nuclear molding, cell–cell wrapping.
   D. Desmoplastic/nodular variant is characterized by pale, reticulin-free nodules surrounded by reticulin-positive collagen fibers.
   E. Medulloblastoma with extensive nodularity (MBEN) usually has small reticulin-free zones that are enriched with neuropil-like tissue.

✅ **Answer E**

 — MBEN is closely related to desmoplastic/nodular variant but differs in that the reticulin-free zones become large and enriched with neuropil-like tissue, resembling the streaming pattern of central neurocytoma.

❓ 5. **Molecular subtypes of MB, the FALSE answer is:**
   A. Are WNT/Wingless (WNT), Sonic Hedgehog (SHH), Group 3 and Group 4.
   B. Are of prognostic value.
   C. WNT MB arise from cerebellar granule cells.
   D. SHH MB arise from cerebellar granule cells or from neural stem cells.
   E. Group 3 and Group 4 MB are derived from cerebellar granule cell precursors.

✅ **Answer C**

 — WNT MBs are more likely to arise from the dorsal brainstem or ventricular zone progenitors early in embryonal development.

❓ 6. **WNT MBs, the FALSE answer is:**
   A. Occur in adults and less frequently in older children.
   B. Male-to-female ratio 1:1.
   C. Rarely metastasize with very good prognosis.
   D. Histologically classic rarely large cell/anaplastic.
   E. Can be associated with monosomy 6 or Turcot's syndrome.

✅ **Answer A**

 — WNT tumors occur in older children and less frequently in adults but are not typically observed in children age 3 years or younger.

❓ 7. **SHH MB, the FALSE answer is:**
   A. More prevalent in children less than 4 years of age and adults.
   B. Male-to-female ratio 1:1.
   C. Histologically, desmoplastic/nodular, classic, and large cell/anaplastic.
   D. Commonly metastasize with very poor prognosis.
   E. Can be associated with chromosome 10q deletions or Gorlin's syndrome.

✅ **Answer D**

 — Infants in SHH MB typically have a better prognosis than older children or adults. Metastases are uncommon.

**? 8. Group 3 of molecular MB subgroups, the FALSE answer is:**
  A. Peak incidence at age 4–6 years.
  B. More common in males (male-to-female ratio $\geq$2:1).
  C. Worst prognosis among the four subgroups.
  D. Infrequently metastasize.
  E. Histologically, majority are classic but exhibit large cell/anaplastic more frequently than any other subgroup.

**✓ Answer D**
  — Metastases are present at the time of diagnosis in up to half of patients with Group 3 MBs, with higher rates of metastatic disease in infants compared to older children.

**? 9. Group 4 of molecular MB subgroups, the FALSE answer is:**
  A. Least common tumors of the four subgroups.
  B. Male-to-female ratio of 3:1.
  C. Peaking in middle to late childhood.
  D. Females frequently exhibit X chromosome deletions.
  E. Intermediate prognosis with adults faring worse than children.

**✓ Answer A**
  — Group 4 MB are the most common tumors of the four subgroups (34–36% of MB). Typically, exhibit classic histology, but desmoplastic/nodular and large cell/anaplastic histology has been observed.

**? 10. Clinical presentations of MB, the FALSE answer is:**
  A. Often due to raised ICP.
  B. Most common signs are papilledema, truncal and limb ataxia and nystagmus.
  C. WNT and Group 4 tumors were observed to have longer median prediagnostic intervals (8 weeks).
  D. Prediagnostic intervals for SHH and Group 3 tumors are 2 and 4 weeks, respectively.
  E. Obstructive hydrocephalus may not be as frequent in pediatric as it is in adults.

**✓ Answer E**
  — Clinical presentation with obstructive hydrocephalus may not be as frequent in adult cases of MB as in pediatric cases younger age also was found to be significantly associated with longer prediagnostic interval although a longer prediagnostic interval of symptoms has been suggested to signify a better prognosis.

**? 11. Imaging of MB, the FALSE answer is**
   A. Predominantly (vermian) in children and more frequently lateral or hemispheric in adults.
   B. Hypointense on T1-weighted MRI but variable on T2-weighted MRI.
   C. Homogeneous contrast-enhancement is frequent in adult.
   D. Lateral location in adults is related to underlying desmoplastic/nodular histology.
   E. Complete spinal MRI should be obtained preoperatively to assess for leptomeningeal dissemination.

**✓ Answer C**
- The intense or homogeneous contrast-enhancement pattern that is frequent in children may not be as common in adults. For the purposes of staging and directing adjuvant therapy, postoperative MRI should be obtained within 48 h after surgery for the assessment of residual tumor.

**? 12. Commonest tumor associated with drop metastasis**
   A. MB.
   B. Ependymoma.
   C. Brainstem glioma.
   D. Choroid plexus papilloma.
   E. Meningioma.

**✓ Answer A**

**? 13. Poor prognostic factors of MB, the FALSE answer is:**
   A. Age 3 years or younger.
   B. Leptomeningeal dissemination.
   C. Residual tumor ($\geq 1.5$ cm$^2$) following neurosurgical resection.
   D. Desmoplastic/nodular and MBEN.
   E. Anaplastic/large cell MB.

**✓ Answer D**
- Best outcomes typically in patients with desmoplastic/nodular and MBEN tumors, intermediate outcomes in patients with classic histology.

**? 14. Staging system for MB, the FALSE answer is:**
   A. T1 tumor <3 cm in diameter.
   B. T2 tumor $\geq 3$ cm in diameter.
   C. T3b tumor >3 cm in diameter with extension.

D. M1 tumor cells found in CSF.
E. M2 intracranial tumor beyond primary site.

✓ **Answer C**

– For Chang et al. staging system for MBs check ◻ Table 10.1.

❓ **15. Adjuvant therapy for MB, the FALSE answer is:**
A. Total irradiation is 64 Gy.
B. Usual chemotherapy protocol: lomustine, vincristine, and cisplatin.
C. Peripheral neuropathy and hematotoxicity may be observed with chemotherapy.
D. Early-onset puberty and hypothyroidism may occur after radiotherapy.
E. Children younger than 3 years are rarely treated with craniospinal irradiation.

✓ **Answer A**

– Radiation therapy in adult MB typically consists of craniospinal radiation with doses of 36 Gy and a boost of 18 Gy to the posterior fossa (thus a total of at least 54 Gy to the posterior fossa).

❓ **16. Ependymomas, the FALSE answer is:**
A. Rare tumors of the CNS.
B. Arise from ependymal cells lining ventricles of brain or central canal of spinal cord.
C. More common in adult than pediatrics.
D. Most common location is infratentorial compartment.
E. In the supratentorial compartment, they arise from nests of ependymal cells that have migrated from periventricular areas.

◻ **Table 10.1** Chang et al. staging system for MBs

T1 tumor <3 cm in diameter
T2 tumor ≥3 cm in diameter
T3a tumor >3 cm in diameter with extension
T3b tumor >3 cm in diameter with unequivocal extension into the brainstem
T4 tumor >3 cm in diameter with extension up past the aqueduct of Sylvius or down past the foramen magnum (i.e., beyond the posterior fossa) or both
M0 no evidence of subarachnoid or hematogenous metastasis M1 tumor cells found in CSF
M2 intracranial tumor beyond primary site
M3 gross nodular seeding in spinal subarachnoid space
M4 metastasis outside the cerebrospinal axis (especially to bone marrow, bone)

*M* metastasis, *T* tumor size and invasion

✅ **Answer C**
- Constitute 3–5% of CNS tumors in adults and nearly 10% of CNS tumors in children.

❓ **17. Pathology of ependymoma, the FALSE answer is:**
   A. Soft gray or tan lesions with well-demarcated borders.
   B. May contain cystic, hemorrhagic, necrotic, or calcified regions.
   C. The presence of perivascular pseudo-rosettes and ependymal rosettes.
   D. WHO I (subependymoma and myxopapillary ependymoma).
   E. WHO II (ependymoma and anaplastic ependymoma).

✅ **Answer E**
- WHO grade II (ependymoma) and grade III (anaplastic ependymoma).

❓ **18. Histopathologic characteristics of ependymomas, the FALSE answer is:**
   A. Subependymoma: clusters of cells with largely isomorphic nuclei with predilection to lateral ventricle.
   B. Myxopapillary ependymoma: cuboidal or spindle-shaped glial cells.
   C. Tanycytic ependymoma: elongated, spindle-like bipolar cells.
   D. Clear cell ependymoma: sheets of uniform cells with centrally located round nuclei.
   E. Papillary ependymoma: cuboidal or columnar epithelial cells resting on fibrillary glial stroma.

✅ **Answer A**
- Subependymoma are clusters of cells with largely isomorphic nuclei with predilection to the fourth ventricle.

❓ **19. Molecular subtypes of infratentorial ependymomas, the FALSE answer is:**
   A. It has been better characterized than that of supratentorial ependymomas.
   B. Group A occurs more frequently in infants.
   C. Group A is located centrally near the pons and less aggressive.
   D. Group B occurs more frequently in adolescents and adults.
   E. Group B demonstrates a high degree of chromosomal defects.

✅ **Answer C**
- Group A is located laterally in the cerebellum and more aggressive.

**? 20. Hereditary tumor syndromes with ependymomas, the FALSE answer is:**
A. The most frequent chromosomal defects seen involves chromosome 22. Neurofibromin 2.
B. Li–Fraumeni syndrome is due to germline mutation in p53 tumor suppressor gene.
C. If associated with NF2, the ependymomas are mostly located in posterior cranial fossa.
D. May be associated with multiple endocrine neoplasia type 1 syndrome.
E. Reported with Turcot's syndrome.

**✓ Answer A**
— There is an increased incidence of intramedullary spinal ependymomas in patients with NF2, but intracranial ependymomas do not harbor the causative mutation in the *NF2* gene on chromosome 22q.

**? 21. Clinical features of ependymoma, the FALSE answer is:**
A. In pediatrics, mean age at diagnosis is 1–3 years.
B. 5–20% of patients present with leptomeningeal dissemination at the time of diagnosis.
C. Infratentorially present with hydrocephalus and cranial nerve deficits.
D. Lower grade tumors have a slower, more gradual onset of symptoms.
E. The median time to recurrence is 22–25 months.

**✓ Answer A**
— In pediatrics, mean age at diagnosis is 5–8 years.

**? 22. Prognostic factors of ependymoma, the FALSE answer is:**
A. Extent of disease at diagnosis.
B. Age of the patient at diagnosis.
C. Tumor location.
D. Low score in grading system.
E. Extent of surgical resection.

**✓ Answer D**
— Current tumor grading systems do not appear to have major prognostic implications, but more recent molecular classification does influence outcomes.

**? 23. Imaging of ependymoma, the FALSE answer is:**
A. Supratentorially, it has greater cystic areas compared to infratentorially.
B. CT scans may be isodense or hyperdense.

C. Nearly half show calcifications.

D. T1-weighted appears hypointense or isointense and T2-weighted isointense or hyperintense.

E. Usually originated from the roof of fourth ventricle.

✓ **Answer E**

– Infratentorial ependymomas arise from the floor (60%), lateral aspect (30%), or roof (10%) of the fourth ventricle.

? **24. Adjuvant therapy in ependymoma, the FALSE answer is:**

A. Chemotherapy plays an important role.

B. Observation without radiotherapy for supratentorial low-grade tumor after gross total resection.

C. Infratentorially, RT is the current standard of care irrespective of the extent of surgical resection.

D. In anaplastic ependymomas, surgery should be accompanied by adjuvant RT irrespective of the extent of resection or tumor location.

E. Stereotactic radiosurgery may be used for residual and recurrent tumors.

✓ **Answer A**

– No established role for chemotherapy in the primary treatment of ependymomas in adults. In infants, several studies have examined the role of adjuvant chemotherapy without any clear benefit.

? **25. Hemangioblastomas, the FALSE answer is:**

A. Account for 7–12% of posterior fossa tumors.

B. More frequently in female.

C. Occur sporadically (60–75% of cases).

D. Found nearly exclusively in the brainstem, spinal cord, and cerebellum.

E. (WHO) grade I CNS neoplasms.

✓ **Answer B**

– Hemangioblastomas are found more frequently (1.5–2 times) in males than in females.

? **26. Peritumoral cyst formation in hemangioblastoma, the FALSE answer is:**

A. Infrequently associated with hemangioblastoma.

B. Initiated by increased vascular permeability.

C. Extravasation of plasma ultrafiltrate into the tumor interstitial spaces.

D. High interstitial tumor pressure then drives this fluid into the surrounding parenchyma.

E. Peritumoral edema forms when the resorptive capacity of the peritumoral tissue is exceeded.

✅ **Answer A**

— About 80% of CNS hemangioblastomas are associated with a peritumoral cyst (a cyst that forms at the margin of the tumor).

❓ **27. Pathology of hemangioblastoma, the FALSE answer is:**

A. Well-circumscribed, encapsulated tumors.
B. Bright red or red-orange mass.
C. Are invariably associated with intense vascularity.
D. Characterized by proliferation of stromal cells and endothelial cells.
E. Histologically similar to squamous cell carcinoma.

✅ **Answer E**

— Because the stromal cells have numerous lipid-containing vacuoles that have a clear cell morphology similar to that of renal cell carcinoma.

❓ **28. Von Hippel-Lindau disease (VHL), the FALSE answer is:**

A. Familial neoplasia syndrome with autosomal dominant mode of inheritance.
B. Hemangioblastoma is the most common CNS manifestation.
C. Epididymal cystadenoma is the major cause of death in VHL.
D. May associated with pheochromocytomas and pancreatic neuroendocrine tumors.
E. In pediatrics, ophthalmoscopy is performed yearly to exclude retinal hemangioblastomas.

✅ **Answer C**

— Primary cause of VHL disease-related death is renal cell carcinoma or CNS hemangioblastoma.

❓ **29. Imaging of hemangioblastoma, the FALSE answer is:**

A. Vividly and discretely enhance on T1-weighted MRI.
B. Small tumors can't be detected in MRI.
C. Peritumoral edema and cysts are best detected and monitored with fluid attenuation inversion recovery (FLAIR).
D. Prominent flow voids are often identified with larger hemangioblastomas on T1- and T2-weighted MRI.
E. Preoperative embolization carries a risk of hemorrhage.

✅ **Answer B**

— Even small tumors (2–3 mm in diameter) can be detected and precisely measured by MRI.

**30. Clinical manifestations and locations of hemangioblastomas, the FALSE answer is:**
  A. Cerebellar hemangioblastomas frequently occur in posterior and medial aspect of cerebellum.
  B. Supratentorially, they most frequently originate in corpus callosum.
  C. Majority of symptoms caused by peritumoral cyst.
  D. Spinal cord hemangioblastomas usually (96%) arise posterior to dentate ligament at dorsal root entry zone.
  E. Gait ataxia, dysmetria, hydrocephalus, and hypesthesia are common.

**Answer B**
  - When hemangioblastomas are found supratentorially, they most frequently (30%) originate in the region of the pituitary stalk and tuber cinereum.

**31. Management of hemangioblastoma, the FALSE answer is:**
  A. Sporadic one treated similar to VHL-related hemangioblastoma.
  B. Microsurgical resection is curative.
  C. Cyst walls are left undisturbed because they are not neoplastic.
  D. Small tumors that are not associated with peritumoral cysts are likely to respond best to radiosurgery.
  E. Chemotherapy has no role in management.

**Answer A**
  - The management strategy for hemangioblastomas in patients with sporadic tumors can differ from the strategy in patients with VHL disease because most patients with VHL disease have multiple CNS hemangioblastomas and because the growth rate of these tumors is unpredictable, treatment of individual tumors is generally postponed until they become symptomatic.

**32. Epidemiology of cerebellar astrocytomas, the FALSE answer is:**
  A. Account for 15–25% of all childhood CNS tumors.
  B. Mean age at diagnosis of 7 years.
  C. Commonly in female.
  D. Pilocytic astrocytomas is the most common type.
  E. Worldwide incidence of 4.8 per million per year.

**Answer C**
  - An equal male-to-female distribution.

**33. Pathology of pilocytic cerebellar astrocytomas, the FALSE answer is:**
   A. Typically, WHO grade I.
   B. Biphasic pattern of dense fibrillary and microcystic.
   C. Presence of eosinophilic granular bodies.
   D. Rosenthal fibers are pathognomonic feature.
   E. Occasionally contain areas of microvascular proliferation.

**Answer D**
   – Rosenthal fibers found not only in pilocytic astrocytoma but are also seen in craniopharyngiomas and Alexander's disease.

**34. Pathology of fibrillary (diffuse) cerebellar astrocytomas, the FALSE answer is:**
   A. Typically WHO grade II.
   B. More nuclear aneuploidy/tetraploidy.
   C. Cellular atypia.
   D. Abundant of Rosenthal fibers and eosinophilic granular bodies.
   E. The rate of 5-year survival with these lesions is 87.5%.

**Answer D**
   – There are no Rosenthal fibers or eosinophilic granular bodies.

**35. Cerebellar astrocytoma imaging, the FALSE answer is:**
   A. Round, well-circumscribed appearance.
   B. On CT scans appear isodense or hypodense.
   C. Calcification is seen only in a minority of cases.
   D. A mural nodule surrounded by a cyst is seen in majority of cases.
   E. Can appear as solid mass or a complete cyst that can be confused with an arachnoid cyst.

**Answer D**
   – A mural nodule surrounded by a cyst is seen in fewer than half of cases.

**36. Cerebellar astrocytoma imaging, the FALSE answer is:**
   A. Solid mural nodule appears hypointense on T1-weighted MRI.
   B. Not typically generate a large amount of vasogenic edema.
   C. Only the solid component display enhancement.
   D. MRI spectroscopy has a characteristic elevated ratio of choline to $N$-acetyl aspartate.
   E. High apparent diffusion coefficient values in low-grade tumors.

✓ **Answer C**

- The cyst wall and the solid component often display enhancement on both CT scans and MRI. The cyst wall composition on imaging is an important variant because it guides the aggressiveness of therapy. Cysts with only a thin rim of enhancement can often be safely left behind during surgical resection, whereas thicker and more irregular cyst wall enhancement necessitates complete resection.

? **37. Adjuvant therapy in cerebellar astrocytomas, the FALSE answer is:**
A. Is mandatory when gross total resection is not feasible because of brainstem or cerebellar peduncle invasion.
B. Chemotherapy is typically used to delay or avoid radiotherapy.
C. Initial chemotherapeutic regimens are a combination of carboplatin and vincristine.
D. Patients with recurrent disease or leptomeningeal spread, aggressive chemotherapy, followed by radiation.
E. Whole-brain radiation in young children has been linked to decreased IQ.

✓ **Answer A**

- Radiation therapy or chemotherapy is not indicated initially because it has not been shown to delay progression or improve survival, even with WHO grade II tumors, spontaneous regression occurs in one third of patients after subtotal resection. Therefore, initial conservative management with serial surveillance MRI scans until there is evidence of tumor progression. If feasible, repeat surgery may be considered at the time of recurrence; after a second subtotal resection, approximately 70% of cases show spontaneous regression.

✓ **38. Outcome and surveillance in cerebellar astrocytoma, the FALSE answer is:**
A. The risk of local recurrence is 50% after gross total resection.
B. Repeat imaging after surgery is generally obtained at 3-, 6-, 9-, and 18-month intervals.
C. Higher incidence of deficits in balance and coordination.
D. A decreased in IQ may occur.
E. Nearly 40% of patients required CSF diversion.

✓ **Answer A**

- The risk of local recurrence is 10–20% after gross total resection and 50–80% after subtotal resection.

**? 39. Epidermoid cysts, the FALSE answer is:**
   A. Benign lesions that may arise in the spine or intracranially.
   B. May be intradural (usually extra-axial) or extradural.
   C. Usually located near the roof of the fourth ventricle.
   D. Are derived from ectopic inclusions of epithelial cells during neural tube closure.
   E. Represent as malformations of surface ectoderm.

**✓ Answer C**
   – Epidermoid cysts usually occur in the cerebellopontine angle or in the parasellar cisterns.

**? 40. Epidermoid cysts imaging, the FALSE answer is:**
   A. In CT scan, it appears only as hypodense mass.
   B. No surrounding edema.
   C. In T1 and T2 MRI similar intensity of CSF.
   D. Diffusion-weighted is used to differentiate it from arachnoid cyst.
   E. Hydrocephalus is rare.

**✓ Answer A**
   – Typical CT findings are those of a homogeneous nonenhancing hypodense lesion in the subarachnoid space. Occasionally, epidermoids present as high-density masses ("white epidermoids") on CT scans, making diagnosis difficult.

**? 41. Pathology of epidermoid cysts, the FALSE answer is:**
   A. A thin capsule of stratified, keratinized squamous epithelium.
   B. Contains an accumulation of desquamated epithelial cells.
   C. Presence of keratin and cholesterol.
   D. Malignant transformation is rare.
   E. No immunohistochemical staining for epidermoids yet.

**✓ Answer E**
   – Immunohistochemical staining for the carbohydrate antigen CA19–9 has had positive results in intracranial epidermoids. This tumor marker, originally developed as a colon cancer antigen, is also detected in the serum of patients with epidermoid tumors.

**? 42. Treatment of epidermoid cyst, the FALSE answer is:**
   A. Surgery is the modality of choice.
   B. Well-demarcated smooth hypovascular capsule with characteristic pearly flakes.

   C. Primary intracapsular debulking and subsequent removal of the capsule.

   D. Fragments of capsule adherent to important structures are left when necessary to avoid neural and vascular injury.

   E. Intraoperative spillage of cyst contents usually caused bacterial meningitis.

✓ **Answer E**

− Chemical meningitis is associated with intraoperative spillage of cyst contents or preoperative cyst rupture. This event has also been reported as radiographic finding of small fat globules in the subarachnoid and intraventricular spaces. Steroid administration is useful in the treatment of meningitis.

? **43. Exophytic brainstem glioma, the FALSE answer is:**

   A. Arise from the subependymal glia in the floor of the fourth ventricle.

   B. Grows anteriorly and rarely extending into the fourth ventricle.

   C. Younger patients typically have a long history of symptoms such as failure to thrive and vomiting.

   D. Older patients, signs, and symptoms of elevated ICP.

   E. Torticollis is also frequently present as a result of chronic tonsillar herniation.

✓ **Answer B**

− Grow posteriorly, often extending into the fourth ventricle, where the bulk of tumor growth occurs.

? **44. Cervicomedullary brainstem glioma, the FALSE answer is:**

   A. Arise within the intramedullary cervical spinal cord.

   B. Grow dorsally towards the obex of the fourth ventricle.

   C. Hydrocephalus can result due to cerebral aqueduct obstruction.

   D. Tumors based in the medulla often manifest with choking and dysphagia.

   E. Respiratory center involvement can lead to central apnea.

✓ **Answer C**

− Hydrocephalus can result with dorsal obstruction of the fourth ventricle and is accompanied by the typical symptoms of elevated ICP.

**45. Diffuse intrinsic pontine glioma, the FALSE answer is:**
    A.  Cranial nerves VI and VII are the most conspicuously involved.
    B.  Cranial nerve signs are typically bilateral.
    C.  Involvement of cerebellar peduncles can result in either truncal or appendicular ataxia.
    D.  Commonly obstruct the fourth ventricle at the time of diagnosis.
    E.  Engulfment of the basilar artery can occur.

**Answer D**
 - These gliomas often result in diffuse enlargement of the pons and, although large, often do not obstruct the fourth ventricle at the time of diagnosis.

**10**

# Fourth Ventricle: Pathology—Non-Tumors

*Rupesh Raut, Mohamed M. Arnaout, Ali M. Neamah, Saja A. Albanaa, Noor M. Akar, and Mustafa Ismail*

**?** 1.  Blake's Pouch cyst
    **Pathology, the FALSE answer is:**
    A.  It is also known as the rudimental fourth ventricular tela choroidea.
    B.  It represents persistence with expansion of Blake's Pouch.
    C.  It normally regresses during the fifth to eighth gestational weeks.
    D.  Persistence is due to failed perforation of the foramen of Luschka.
    E.  There is posterior ballooning of the inferior medullary velum.

**✓ Answer D**
    - It is caused by a failure of the regression of Blake's Pouch secondary to the non-perforation of the foramen of Magendie.

**?** 2.  Blake's Pouch cyst
    **Pathology, the FALSE answer is:**
    A.  During embryogenesis, foramina of Luschka opens later than Magendie.
    B.  Foramina of Luschka tries to compensate for the CSF outflow to the cisterns.
    C.  The cyst balloons posteriorly into the cisterna magna.
    D.  Cerebellar vermis is well developed.
    E.  The cyst does not communicate with the fourth ventricle.

**?** **Answer E**
    - Blake's Pouch cyst communicates with the fourth ventricle.

**?** 3.  Blake's Pouch cyst.
    **Clinical Presentation, the FALSE answer is:**
    A.  It is a rare entity seen among young age patients.
    B.  It was previously classified as a part of the Dandy–Walker continuum.
    C.  Hydrocephalus is not a feature of this condition.
    D.  Some remain asymptomatic for the rest of their lives.
    E.  CSF shunting has good outcome.

**✓ Answer C**
    - When symptomatic, patients present with impaired neurological symptoms and progressive hydrocephalus with symptoms such as headache, vomiting, blurred or double vision.

**? 4. Blake's Pouch cyst**
   **Radiological features, the FALSE answer is:**
   A. Triventricular hydrocephalus.
   B. Infra- or retrocerebellar localization of the cyst.
   C. Well-developed nonrotated cerebellar vermis.
   D. Cystic dilation of the fourth ventricle.
   E. Some degree of compression on the medial cerebellar hemi-
      spheres.

**✓ Answer A**
   – Hydrocephalus usually involves the fourth ventricle and supratentorial ventricles producing classical tetra-ventricular hydrocephalus.
   Several articles stress on the presence of a tetraventricular hydrocephalus to make the diagnosis.

**? 5. Mega cisterna magna, the FALSE answer is:**
   A. It is a normal variant, occurs in ~1% of all brains imaged.
   B. It is a focal enlargement of the CSF-filled subarachnoid space.
   C. Cerebellar atrophy is a common finding.
   D. All the children undergo normal development.
   E. It communicates with the fourth ventricle.

**✓ Answer D**
   – It is an incidental finding on neuroimaging, and no imaging follow-up
   is necessary as patients are usually asymptomatic. But there are now
   studies proving that it may affect movement ability, as well as adapt-
   ing, social behavior, and language abilities during child development.

**? 6. Arachnoid cyst.**
   **The FALSE answer is:**
   A. Arachnoid cysts are benign developmental anomalies.
   B. It communicates with adjacent ventricles and cisterns.
   C. Occur in virtually all locations where arachnoid is present.
   D. These are formed by splitting or duplication of the arachnoid
      membrane during the formation of the subarachnoid cisterns.
   E. Commonly located in or adjacent to the Sylvian fissure, CP
      angle.

**✓ Answer B**
   – Arachnoid cyst normally does not communicate with the cisterns, ventricles. These are filled with CSF and exert pressure over the adjacent
   structures.

**7. Arachnoid cyst.**
**The FALSE answer is:**
A. Hydrocephalus is a common finding.
B. Symptoms are typically due to progressive brainstem compression.
C. Definitive treatment consisted of ventriculoperitoneal shunt.
D. Endoscopic fenestration is another treatment modality.
E. MRI flowmetry helps to delineate the flow of CSF, differentiates it from trapped fourth ventricle.

**Answer C**
- The definitive treatment for an arachnoid cyst in the fourth ventricle is complete surgical excision of the cyst via a median suboccipital approach.

**8. Trapped fourth ventricle.**
**The FALSE answer is:**
A. Observed in children after ETV for the treatment of hydrocephalus.
B. Blockage of both the outlets (Luschka and Magendie) and the inlet of the fourth ventricle at the level of the cerebral aqueduct.
C. Pressure in the fourth ventricle increases due to abnormal flow of CSF.
D. Brainstem compression and lower cranial nerve palsies.
E. Treatment includes shunts, endoscopic surgery, and foramen magnum decompression by fenestration of the trapped fourth ventricle.

**Answer A**
- The trapped fourth ventricle is a rare late complication seen in children after CSF diversion for the treatment of postinfectious or posthemorrhagic hydrocephalus after with lateral ventricle shunting. Functional occlusion of the aqueduct is often related to over drainage with or without associated slit ventricles, and the occlusion of foramina is due to arachnoiditis secondary to infection or hemorrhage.

**9. Dandy–Walker malformation.**
**Pathophysiology, the FALSE answer is:**
A. Most common posterior fossa malformation with its enlargement.
B. Partial (hypoplasia) or complete agenesis of the cerebellar vermis.

C. Cystic dilatation of the fourth ventricle which is distorted.

D. The fourth ventricle is encased in a neuroglial vascular membrane.

E. Herniation of cerebellar tonsils >5 mm below the foramen magnum.

✅ **Answer E**

- Herniation of cerebella tonsils is associated with Chiari malformation and not Dandy–Walker malformation.

❓ **10. Dandy–Walker malformation.**
**Clinical presentation, the FALSE answer is:**

A. Most patients present in their first year of life.

B. The most common manifestation is macrocephaly.

C. Patients may be syndromic with malformations of the heart, face, limbs, and gastrointestinal or genitourinary system.

D. 90% of patients have some degree of mental retardation.

E. Hydrocephalus occurs in 70–90% of these cases.

✅ **Answer D**

- 50% of patients with Dandy–Walker malformation have normal IQ and the rest of the 40–50% have some degree of mental retardation.

❓ **11. Dandy–Walker malformation.**
**MRI findings, the FALSE answer is:**

A. Atresia of Magendie and Luschka.

B. Tentorium and torcular herophili are usually displaced downward.

C. Agenesis of the cerebellar vermis.

D. Enlarged posterior fossa cyst that communicates with the fourth ventricle.

E. Possible agenesis of the corpus callosum.

✅ **Answer B**

- The DWM is a heterogeneous disorder characterized by hypoplasia and upward rotation of the cerebellar vermis, cystic dilation of the fourth ventricle, and an enlarged posterior fossa with upward displacement of the lateral sinuses, tentorium, and torcular heterophili.

**? 12. Dandy–Walker malformation.**
**Treatment, the FALSE answer is:**
A. In the absence of symptoms, DWM may be managed by follow-up only.
B. When symptomatic, ETV is an option where the aqueduct is patent.
C. Posterior fossa craniotomy with excision of obstructing membrane.
D. Ventriculoperitoneal (VP) shunt alone.
E. VP with cystoperitoneal shunting.

**✓ Answer D**
— Shunting the lateral ventricles alone is contraindicated because of the risk of upward herniation with increased pressure in the posterior fossa.

**? 13. Dandy–Walker malformation.**
**Prognosis, the FALSE answer is:**
A. Despite the intervention, there are 12–50% mortality rates.
B. Prognosis largely depends upon the hydrocephalus, congenital abnormalities, and brain malformations.
C. The risk of seizure is 90–100%.
D. Fifty percent of children with untreated hydrocephalus die before age 3.
E. Around 20% will reach adult life.

**✓ Answer C**
— The risk of epileptic seizures ranges from 15 to 30%.

**? 14. Dandy–Walker Variant (DWV).**
**Characteristics, the FALSE answer is:**
A. Less severe anomaly than Dandy–Walker malformation.
B. Mild vermian hypoplasia.
C. A small posterior fossa.
D. A small cystic cavity that communicates with the fourth ventricle.
E. Hydrocephalus is present (<30%) but not as commonly as in DWM.

**✓ Answer C**
— The posterior fossa in DWV is of normal size. It is enlarged in DWM. Cystic cavities communicate with the fourth ventricle in both DMW and DWV.

**?** **15. Epidermoid cyst.**

**Clinical presentation, the FALSE answer is:**

A. It compromises 0.2–1.8% of all primary intracranial lesions.

B. Less than 5% occur within the fourth ventricle.

C. Symptoms occur due to cerebellar compression.

D. Symptoms can include headaches, double vision, facial palsy, gait ataxia, hearing impairment, trigeminal neuralgia, and facial tics.

E. Patients often present in the first decade of life.

**✓ Answer E**

— Association in the pediatric age is very unusual. These tumors grow very slowly by the gradual accumulation of normally dividing cells, and often sufficient time is required to reach a size large enough to cause clinical symptoms. Patients often present in the fourth decade of life.

**?** **16. Epidermoid cyst.**

**Investigations, the FALSE answer is:**

A. MRI shows low signal intensity on T1WI and high on T2WI.

B. There is homogenous enhancement after gadolinium on T1WI.

C. CT shows homogeneously hypodense, circumscribed areas with no edema, no calcification.

D. The margin of the tumor is irregular with a "flowing" pattern of growth, no peritumoral edema and hydrocephalus.

E. There is abnormal restricted diffusion in DWI.

**✓ Answer B**

— Epidermoid cysts do not show contrast enhancement in CT or MRI. They appear bright as they show restricted diffusion in DWI and thus, DWI helps differentiate them from the arachnoid cysts.

**?** **17. Epidermoid cyst.**

**Treatment, the FALSE answer is:**

A. Complete total excision is always possible.

B. If the germinal capsule is adherent to the floor of the fourth ventricle, it is not excised.

C. The remnants should be devitalized to avoid spillage of keratinous material.

D. Intra-operative dispersion of cyst contents should be avoided to reduce the risk of aseptic meningitis.

E. Due to slow growth rate, it takes years to recur to a significant size.

✅ **Answer A**
- Epidermoid cysts are very notorious as they grow around neurovascular tissues and sometimes are adherent to the floor of the fourth ventricle. In such cases, the capsule is left untouched to avoid morbidity and mortality. The goal here is to do the maximum safe resection.

❓ **18. Neurocysticercosis.**

**Characteristics, the FALSE answer is:**
A. Should raise suspicion in certain geographic locations.
B. Often presents in multiple ventricular and parenchymal sites.
C. Cysts are mobile, tend to migrate within the ventricular system from time to time.
D. Among the ventricles, the third ventricle is the most common site of its occurrence.
E. Two types of cysticercosis are described-cellulosa (cyst with scolex) and racemose (cyst lacks scolex).

✅ **Answer D**
- The fourth ventricle is said to be the favored site of intraventricular neurocysticercosis, probably due to the gravitational effect that favors the migration of the cysts from the superior cavities. It results in entrapment of the cysts within the fourth ventricle due to the small size of the outlet foraminae and can produce acute obstructive hydrocephalus and death in patients.

❓ **19. Neurocysticercosis.**

**Investigations, the FALSE answer is:**
A. Immunological studies are neither sensitive nor specific.
B. Contrast-enhanced CT may not show enhancement.
C. MRI is the investigation of choice.
D. CSF flow study shows the intraluminal nature of the cyst.
E. Contrast-enhanced T1WI may not show enhancement.

✅ **Answer E**
- Contrast-enhanced CT may not show enhancement, whereas the contrast-enhanced MRI exhibits enhancement, suggesting that MRI is more sensitive in detecting underlying ependymitis. The 3D-CISS-MRI sequence is used for the demonstration of an intraventricular cysticercus cyst—the scolex, cyst wall, and cyst fluid.

# Fourth Ventricle: Surgery

*Waeel O. Hamouda, Yara Alfawares,
Vishan P. Ramanathan, and Mustafa Ismail*

**?** 1. **Preoperative treatment with steroids.**
   **All can occur. False answer is:**
   A. Decreases vasogenic edema.
   B. Decreases headache and neck pain.
   C. Decreases the incidence and severity of aseptic meningitis.
   D. Increases the incidence of posterior fossa syndrome.
   E. Decreases nausea and vomiting allowing for better hydration and nutrition prior to surgery.

**✓ Answer D**
   − Steroids decrease the incidence of posterior fossa syndrome.

**?** 2. **Intraoperative monitoring of the brainstem structures at the fourth ventricle floor.**
   **Methods. False answer is:**
   A. Vital signs (HR and BP) monitoring is not a part of intraoperative neurophysiological status.
   B. Brainstem auditory-evoked potentials.
   C. Somatosensory-evoked potentials (SSEP).
   D. Free running electromyography.
   E. Transcranial motor-evoked potential.

**✓ Answer A**

**?** 3. **Intraoperative monitoring of the brainstem structures at the fourth ventricle floor.**
   **Methods. False answer is:**
   A. HR and BP signal violation of sympathetic nuclei in brainstem.
   B. BAEP signals violation of cochlear nucleus, superior olive, or lateral lemniscus.
   C. SSEP signals violation of medial lemniscus.
   D. Electromyography signals violation of VII, IX, X, and XII nuclei and nerves.
   E. Transcranial motor-evoked potential for corticospinal tracts and cranial nerves.

**✓ Answer A**
   − HR and BP signal violations in nucleus tractus solitaries and dorsal motor nucleus of the vagus nerve in medulla.

**? 4. Intraoperative monitoring of brainstem at fourth ventricle floor. Limitations. False answer is:**

A. HR and BP monitoring is the least sensitive measure of alteration of brainstem function.

B. BAEP pathway may be intact despite damage to central brainstem.

C. SSEP pathway is deep away from the floor of the fourth ventricle.

D. Electromyography might need direct stimulation of tested nerves or nuclei for accurate mapping.

E. Transcranial motor-evoked potential can elicit fine jerks of limbs during recording.

**✓ Answer A**

— Heart rate and blood pressure is the most sensitive way to monitor brainstem functions.

**? 5. Positioning during approaches to the fourth ventricle. General consideration. False answer is:**

A. Can be prone, Concord prone, three-quarter (lateral oblique) prone, or sitting positions.

B. Prone position is optimal for very young children.

C. All positions require a certain amount of neck extension.

D. Preexisting neck pathologies should be excluded, e.g., craniocervical instability.

E. Head should be translated posteriorly then flexed as much as possible.

**✓ Answer C**

— Certain amount of neck flexion is needed to allow visualization of the rostral part of the fourth ventricle by the surgical microscope and to open up the space between the foramen magnum and the arch of C1.

**? 6. Mayfield head fixation for approaches to the fourth ventricle. General considerations. False answer is:**

A. Must be used with the Concord prone, three-quarter prone and sitting positions.

B. Can be used for patients of all ages.

C. Pins are placed 2 cm above the ear in the unshaven scalp.

D. Pins placement should avoid the squamous temporal bone and shunt tubing if present.

E. Skull penetration by pins can produce depressed fracture, extradural hematoma, or postoperative abscess.

**✓ Answer B**

— Head fixation pins should not be used for patients below the age of 2 years.

**? 7. Prone position for approaching the fourth ventricle.**
**Advantages. False answer is:**
  A. Anatomical orientation is in midline and unrotated.
  B. Allow two surgeons to operate together one on each side of the head.
  C. Avoid usage of fixation pins in very young children by using padded horseshoe.
  D. Excellent visualization of the upper part of the fourth ventricle.
  E. Avoid complications of sitting position, e.g., air embolism.

**✓ Answer D**

— Visualization of pathologies high in the fourth ventricle in prone position is not readily feasible, can be further limited by inadequate neck flexion, and might require a midline vermian incision.

**? 8. Prone position for approaching the fourth ventricle.**
**Disadvantages. False answer is:**
  A. Venous congestion which leads to more bleeding and facial soft tissues swelling.
  B. Central retinal artery or vein thrombosis from pressure on the eyes.
  C. Slippage of the endotracheal tube by drooling saliva.
  D. The posterior fossa contents sink inward.
  E. Constantly requires supporting upper cerebellar hemisphere to maintain exposure.

**✓ Answer E**

— In prone position, both cerebellar hemispheres are on the same level on each side of the midline and require minimal retraction to expose the fourth ventricle. On the contrary, supporting the upper cerebellar hemisphere is mandatory in the lateral park bench or three-quarter prone positions.

**? 9. Prone position for approaching the fourth ventricle.**
**Precautions. False answer is:**
  A. Neck is placed in a neutral position.
  B. Chin and chest should be at least two fingers apart to avoid jugular veins compression.

   C. Table is positioned so that the head is above the heart to improve venous congestion.

   D. Shoulders can be gently pulled toward feet with adhesive tape to stretch nape skin.

   E. Strap under the buttocks to prevent patient sliding in anti-Trendelenburg position.

✅ **Answer A**

— The neck is placed in the "military tuck position" with moderate flexion of the upper cervical spine (to open up the space between the foramen magnum and the arch of C1) and mild extension of the lower cervical spine (to bring the occiput at level parallel with the patient's back).

❓ 10. **Sitting position for approaching the fourth ventricle.**
**Complications. False answer is:**

   A. Cardiovascular instability and hypotension.

   B. Air embolism.

   C. Subdural hematoma.

   D. Median nerve neuropathy.

   E. Tension pneumocephalus.

✅ **Answer D**

— Median nerve neuropathy is not a reported complication of the sitting position. However, cervical myelopathy from neck elevation and flexion can occur. Also, sciatic nerve palsy from prolonged sitting with extended knees might occur.

❓ 11. **Sitting position when approaching the fourth ventricle.**
**Precautions. False answer is:**

   A. Exclusion of right to left shunt through a patent foramen ovale.

   B. Patient is elevated rapidly into sitting position for effective reduction of ICP.

   C. VP shunt should be occluded if possible, prior to positioning.

   D. CSF loss from supratentorial ventricles after a fourth ventricle lesion excision should be allowed slowly.

   E. Vaseline gauze or ointment should be applied at Mayfield pins site.

✅ **Answer B**

— The patient is elevated slowly into the sitting position to avoid sudden orthostatic hypotension in absence of muscle tone under the effect of muscle relaxants.

- VP shunt better to be occluded if possible, prior to attempting an operation in the sitting position to reduce the risk of developing subdural hematoma.
- Vaseline gauze or ointment should be applied at the pins site during applying the head holder to minimize entry of air through penetrated scalp or skull or dural venous channels.

**?** 12. Midline suboccipital approach.
   **Skin incision and suboccipital muscles dissection. False answer is:**
   A. Incision usually extends in midline from 3 cm above the inion to C3 spinous process.
   B. Injury to vertebral artery should be avoided during undermining the skin flaps.
   C. Fascia and underlying muscles are incised in Y-shape to facilitate later closure.
   D. Inferior vertical limb of the Y passes through the avascular ligamentum nuchae.
   E. Superior oblique limbs of the Y extend laterally till close to the mastoid processes.

**✓ Answer B**
   - Injury to the occipital artery and greater occipital nerve should be avoided whenever possible during undermining the skin flaps superficial to the fascia on both sides at the rostral half of the incision, as they traverse the fascia at a point 1–2 cm inferior to the superior nuchal line and 3–4 cm lateral to the midline (nerve medial to the artery) to supply the scalp at the back of the head.

**?** 13. Midline suboccipital approach.
   **Suboccipital craniotomy. The false answer is:**
   A. In children, dura is not adherent to skull, so it is safe to drill close or above the sinuses.
   B. Superior and lateral limits of craniotomy are transverse and sigmoid sinuses, respectively.
   C. Inferior limit of craniotomy includes the posterior edge of the foramen magnum.
   D. The midline keel can be easily removed by simply pulling it using a bone nibbler.
   E. All exposed bone edges should be waxed to avoid air embolism.

✅ **Answer D**
- The midline keel can be very deep and extreme caution must be exercised during stripping the dura off the midline keel using a Penfield, especially near the occipital sinus in the midline and the annular sinus near the foramen magnum to avoid sinuses tears and profuse bleeding.
- Inferior limit of craniotomy should include the posterior edge of the foramen magnum as far laterally as the occipital condyles to allow exposure of the tonsils.

❓ **14. Midline suboccipital approach.**
   **Exposing C1 posterior arch. False answer is:**
   A. Start on the inferior aspect since the vertebral artery lies on its superior aspect.
   B. Monopolar cautery usage to expose the C1 arch might injure an aberrant vertebral artery.
   C. C1 posterior arch can be bifid in infants and young children.
   D. Vertebral artery is found about 25–30 mm of midline on upper aspect of C1 arch.
   E. It is safer to always keep contact of the periosteal elevator with the underlying bone.

✅ **Answer D**
- The vertebral artery lies an average of 15 mm from the midline on the upper aspect of C1 posterior arch.

❓ **15. Midline suboccipital approach.**
   **Upper cervical laminectomy. False answer is:**
   A. C1 laminectomy is needed in certain situations.
   B. Children might use soft cervical collar for 8 weeks until paraspinal muscles reattach.
   C. Laminectomies can be performed safely till C4 in pediatrics.
   D. For tumors, it is only necessary to remove single level below caudal aspect of tumor.
   E. Children do cervical X-rays every few months for few years to check spinal deformities.

✅ **Answer C**
- Extending laminectomies below C2 in young children increase the risk of swan neck deformity.
- C1 laminectomy is needed in for lesions that extend beneath the foramen magnum, in cases associated with tonsillar descend or to permit the surgeon to angle the instruments upward.

**? 16. Midline suboccipital approach.**
**Dural opening. False answer is:**
A. The occipital and annular sinuses are relatively large under the age of 2 years.
B. Y-shaped incision is usually used to allow wide visualization and extension if necessary.
C. Bleeding from occipital or annular sinuses is controlled with clips or sutures.
D. Vertical limb extends downward to foramen magnum below the falx cerebelli.
E. The next step after opening of the dura is direct debulking of the tumor.

**Answer E**
- Once dura is opened, arachnoid over the cisterna magna is opened next to allow CSF drainage and brain relaxation before any attempts for tumor excision.

**? 17. Midline suboccipital approach.**
**Dural closure. False answer is:**
A. Valsalva maneuver can identify potentially unsecured sources of venous bleeding.
B. If dura is not closed watertight, there is high risk of pseudo-meningocele.
C. If dura is not closed watertight, there is low risk of developing post-op hydrocephalus.
D. Bovine pericardium or human allograft dura may produce post-op aseptic meningitis.
E. Shrunken dural defect usually occurs after bipolar obliteration of sinuses.

**Answer C**
- On the contrary, there is increased risk of developing hydrocephalus from arachnoid adhesions produced by blood trickling intradurally from the suboccipital muscles if the dura is not closed watertight.

**? 18. Midline suboccipital approach to fourth ventricle tumors.**
**Telovelar approach. False answer is:**
A. Tonsils freed before its elevation from adhesions to overlying dura and cisterna magna.
B. PICA caudal loops should be identified early being tethered to medial walls of tonsils.

    C. Tonsils and, uvula elevated, and separated to expose PICA in ipsilateral cerebello-medullary fissure.
    D. Exposed tela choroidea is incised to level of the inferior medullary velum and opex.
    E. All PICA branches met during incising the tela choroidea can be coagulated.

✓ **Answer E**
- Only the choroidal branches of the PICA can be coagulated with preservation of the branches supplying the neuraxis.

❓ **19. Midline suboccipital approach to fourth ventricle tumors.**
**Tumor exposure nuances. False answer is:**
    A. Cottonoid patties are not used as tool to dissect tumor from floor of fourth ventricle.
    B. Tela choroidea and inferior medullary velum are opened to expose caudal fourth ventricle.
    C. Retracting or incising caudal vermis can provide exposure to rostral fourth ventricle.
    D. Lesions in lateral roof or lateral recess may require the removal of the ipsilateral tonsil.
    E. Tumor extending through foramina of Luschka is difficult to access.

✓ **Answer E**
- The portion of the tumor extending extra ventricularly through the foramen of Luschka in the cerebello-pontine cisterns can be approached either by retracting the ipsilateral tonsil and cerebellar hemisphere superomedially or by performing a secondary retromastoid approach.
- Lesions in lateral roof or lateral recess may require the removal of the ipsilateral tonsil (by dividing its pedicle attached to the biventral lobule) which may be further augmented by resecting part of the ipsilateral cerebellar hemisphere.

❓ **20. Preoperative hydrocephalus to fourth ventricle tumors.**
**False answer is:**
    A. One of the most significant causes of mortality associated with fourth ventricular tumors.
    B. Papilledema is constant finding with preoperative hydrocephalus especially in children.

    C. Increases perioperative morbidity 2ry to malnutrition from repeated vomiting.
    D. Preoperative high dose steroids can produce improvement in hydrocephalus.
    E. Can be managed temporarily with EVD.

**Answer B**

— Papilledema 2ry to increased ICP is not a constant finding in pediatrics due to the underdeveloped subarachnoid spaces surrounding the optic nerves.

**21. Preoperative shunting to fourth ventricle tumors.**
**False answer is:**
    A. All resulted in extraneural peritoneal metastasis.
    B. Used to treat preoperative hydrocephalus.
    C. Expected to prevent postoperative pseudomeningocele, CSF leak, and meningitis.
    D. Associated with increased incidence of subdural hematoma.
    E. May cause upward transtentorial brainstem herniation and compression.

**Answer A**

— The topic is still controversial. The shunt-induced metastasis is rare. Metastasis related to malignant fourth ventricular tumors has the same incidence as in patients without VP shunts in several studies.

**22. EVD to fourth ventricle tumors.**
**False answer is:**
    A. Placed preoperatively if there is disturbed conscious level 2ry to hydrocephalus.
    B. Allows for controlled CSF drainage preoperatively to prevent upward herniation.
    C. Used postoperatively to clear debris, blood, and air within ventricles.
    D. Has no effect on the incidence of postoperative shunting.
    E. Infection rate may be as high as 10%.

**Answer D**

— Statistically, EVD placement reduces the necessity to place permanent shunts postoperatively.

**23. Shunt dependence in fourth ventricle tumors.**
   **False answer is:**
   A. 50% of patients with fourth ventricular tumors require permanent shunting.
   B. Risk factors include younger age, larger pre-op ventricle size, and extensive tumors.
   C. Permanent hydrocephalus is related to slow-growing tumors.
   D. Add the risks of VP shunt placement to the tumor-specific morbidities.
   E. Perioperative placement of temporary EVD reduce postoperative shunt dependence.

**Answer A**
   - Only 10–15% of patients end up with permanent shunts.
   - Permanent hydrocephalus is related to slow-growing tumors such as astrocytoma since more acute tumors rapidly distend the ventricles over a short period of time not allowing outlet adhesions to form.

**24. Pneumocephalus.**
   **False answer is:**
   A. More common with sitting position than with prone one.
   B. Mild and moderate pneumocephalus requires no specific management.
   C. More common when patients have preoperative hydrocephalus.
   D. Frequently results from overzealous intraoperative drainage of CSF.
   E. Intraventricular air may cause VP shunt malfunction due to airlock.

**Answer B**
   - Mild or moderate pneumocephalus can turn to the symptomatic tension type. The conservative treatment involves placing the patient in the Fowler position of 30°, 100% $O_2$ by fitting mask, avoiding Valsalva maneuver (coughing and sneezing), administering pain medications to prevent straining and antipyretic medications to prevent hyperthermia. Success rate in achieving resorption is about 80–85%.

**25. Tension pneumocephalus.**
   **False answer is:**
   A. If recognized intraoperatively, patient placed in Trendelenburg position and, operative bed irrigated to replace trapped air.
   B. Nitrous oxide as an anesthetic agent might contribute to tension pneumocephalus.

C. If symptomatic postoperatively, treated with small frontal burr hole in supine position.
D. The "Mount Fuji" sign is pathognomonic for symptomatic tension pneumocephalus.
E. It is a neurosurgical emergency.

**Answer D**
- The "Mount Fuji" sign can be found even in patients with trauma.

**26. Postoperative suboccipital pseudomeningocele.**
**False answer is:**
A. Incidence 10–15% after posterior fossa surgery.
B. Solely attributed to postoperative seroma collection.
C. More common in children.
D. Can respond well to serial lumbar punctures.
E. Can put the incision closure under tension and eventually produce CSF leak.

**Answer B**
- May represent also a manifestation of hydrocephalus and may require a CSF diversion procedure.

**27. Aseptic meningitis.**
**Incidence. False answer is:**
A. Also called posterior fossa fever.
B. More common after infratentorial surgeries than supratentorial ones.
C. More with epidermoids or dermoids that rupture intra-operatively.
D. Less likely occurs after resection of astrocytoma or MB.
E. It occurs more commonly in adults.

**Answer E**
- More common in pediatric patients.

**28. Aseptic meningitis.**
**Management. False answer is:**
A. Often self-limiting but occasionally runs a protracted course.
B. Present 3–6 days after surgery with fever, headache, meningeal signs, and irritability.
C. CSF analysis shows increased WBC (mainly mononuclear) and increased proteins.

D. Cannot be differentiated from true bacterial meningitis.
E. Resolves with steroids and serial LPs to remove inflammatory cytokines in CSF.

✅ **Answer D**
- Can be differentiated from true bacterial meningitis by negative cultures, normal CSF glucose level and the absence of progressive decline in conscious level.

❓ **29. Cranial nerves palsies after fourth ventricle surgery.**
**False answer is:**
A. VII and VI nerves are the most affected secondary to injury at the facial colliculus area.
B. XII injury during fourth ventricle surgery has low morbidity.
C. Skew ocular deviation results from periaqueductal grey matter injury.
D. Corneal exposure from VII palsy is treated by artificial tears/tarsorrhaphy or permanently by XII–VII anastomosis.
E. Diplopia from VI palsy is treated by eye patch or permanently by eye muscle surgery.

✅ **Answer B**
- XII injury while less common than facial palsy is much more serious complication as it is usually bilateral since both hypoglossal nuclei are close together in the median raphe. The patient suffers from swallowing apraxia and continuous drooling with high risk for aspiration especially when combined with concomitant facial palsy.

❓ **30. Posterior fossa syndrome.**
**Incidence. False answer is:**
A. Also called posterior fossa mutism, cerebellar mutism, or split vermis syndrome.
B. Occurs almost exclusively in children.
C. Seen in 15% of intraventricular approaches to lesions near the brainstem.
D. Rarely described with supracerebellar infratentorial approach to pineal region and retromastoid approach to side or front of brainstem.
E. Functional prognosis correlates with the resected tumor size.

✅ **Answer E**
- Functional prognosis correlates with initial severity of symptoms as well as the duration of symptoms after surgery.

**31. Posterior fossa syndrome.**
**Clinical picture. False answer is:**
A. Motor deficits in the form of hemiparesis and dependent gait.
B. Mutism with reduction and abnormality in speech production.
C. Behavioral symptoms as withdrawal and internalizing regressive personality.
D. Emotional lability as dysphoria, tearfulness, and apathy.
E. Supranuclear cranial nerves palsies with orofacial apraxia, dysarthria, and dysphagia.

**Answer A**
- There are no actual true motor deficits but rather cerebellar ataxia and hypotonia.

**32. Posterior fossa syndrome.**
**Pathogenesis. False answer is:**
A. Delayed onset of mutism occurring 3–6 days postoperatively.
B. Most patients have some improvement over 3–6 months.
C. Risk factors include vermian incision, large tumors, and preoperative hydrocephalus.
D. Suggested that edema from operative manipulation may play a role.
E. Structures related are vermis, dentate nuclei, and dentato-thalamo-cortical pathways.

**Answer C**
- Risk factors include MB tumors, tumors invading brainstem, and tumors arising from vermis. Other studied factors like vermian incision during surgery, large size tumor, and pre-op hydrocephalus was not constantly proven.
- Because of the delay in onset, it is suggested that edema from operative manipulation may play a role, for example, through transmission of retractor pressure from the medial cerebellum through fiber pathways along the middle and superior cerebellar peduncles into the upper pons and midbrain.

**33. Seizures with fourth ventricles tumors.**
**Perioperative risk factors. False answer is:**
A. Slowly growing tumors.
B. Presence of ventricular drainage or shunting.
C. Subdural hematoma development.

D. Meningitis.

E. Hydrocephalus.

✅ **Answer A**

- Faster growing fourth ventricle tumors are associated with increased risk of symptomatic seizures.

❓ **34. Cerebellar peduncles injury.**

**False answer is:**

A. Middle CP is the most susceptible to injury during surgery.

B. Injury to superior CP produces ipsilateral ataxia and intention tremor.

C. Injury to middle CP causes ataxia and dysmetria.

D. Injury to inferior CP produces dysequilibrium, truncal ataxia, staggering gait, and oscillation of head and trunk on assuming erect position.

E. Injury to inferior CP does not produce ataxia of voluntary movement of the extremities (intension tremors).

✅ **Answer A**

- Since the superior and inferior cerebellar peduncles make up the lateral walls of the superior roof of the fourth ventricle, they are more susceptible to damage during intraventricular procedures than the middle peduncle.

❓ **35. Cerebellar dentate nucleus injury.**

**False answer is:**

A. Lead to ipsilateral limb ataxia, dysmetria, dysdiadochokinesia, and hypotonia.

B. Located at superolateral margin of fourth ventricle roof adjacent to upper pole of tonsil.

C. Most patients have poor recovery after dentate injury.

D. More vulnerable during dissection of hemispheric rather than ventricular tumors.

E. Injury of superior cerebellar peduncle produces similar symptoms.

✅ **Answer C**

- Unless the dentate is completely damaged, most patients recover well within a few months with only minor residual intention tremor that does not interfere with motor development.

**?** **36. Acute urinary retention after fourth ventricle surgery.**
  **False answer is:**
  A. An uncommon complication which is usually reversable.
  B. Respond well to pharmacological therapy.
  C. From injury to pontine micturition center near striae medullaris.
  D. Pontine micturition center integrates cortical motor outputs with sacral sensory inputs that appraise bladder filling status.
  E. Pontine micturition center injury interferes with the ability to initiate voiding in spite of a full bladder.

**✓** **Answer B**
  — The acute urinary retention due to Pontine micturition center injury does not respond to detrusor augmenting agents or alpha-adrenergic blockers but is best managed by intermittent catheterization.

**?** **37. Surgery for MB of the fourth ventricle.**
  **False answer is:**
  A. Patties are placed early along ventricle floor to delineate brainstem posterior surface.
  B. Trajectory should be constantly verified to prevent diving into brainstem at an angle.
  C. Anterosuperior tumor pole is better removed last at the end of operation.
  D. If tumor extends laterally through Luschka, separate retrosigmoid approach is needed.
  E. At the end of surgery, caudal aqueduct opening should be patent with CSF egressing out of it.

**✓** **Answer D**
  — If a small part of tumor extends laterally through the foramen of Luschka, it can be aspirated under direct vision from the ventricular side since the tumor is not adherent to the ependyma.
  — The anterosuperior tumor pole covering caudal opening of the aqueduct is better kept until the end of the operation so that no blood from the dissection will enter the lateral or third ventricles.

**?** **38. Surgery for MB of the fourth ventricle.**
  **False answer is:**
  A. Often immediately visible after the dural opening.
  B. The tumor is usually purple-grey, friable, relatively sackable, and quite vascular.

    C. Desmoplastic tumors cannot be aspirated, being firmer and more fibrous so use Cavitron ultrasonic surgical aspirator.
    D. The tumor is tightly adherent to the brainstem.
    E. Most tumors are attached at the level of middle CP at the lateral recess but rarely midline.

✅ **Answer D**
    — The tumor is rarely adherent or invasive into the brainstem.

❓ **39. Surgery for MB of the fourth ventricle.**
    **Negative prognostic factors. False answer is:**
    A. Larger preoperative tumor size.
    B. Brainstem invasiveness.
    C. Neuraxis dissemination at diagnosis.
    D. Older age of the patient.
    E. Larger postoperative residual tumor.

✅ **Answer D**
    — Children younger than 4 years have worse prognosis than older patients.

❓ **40. Surgery for ependymoma of the fourth ventricle.**
    **False answer is:**
    A. Origin is middle to lower fourth ventricle or at junction of inferior velum with Luschka.
    B. The tumor is usually firm and extensively attached to surrounding structures.
    C. Insinuate into spaces within and about fourth ventricle.
    D. 1-mm thick carpet of tumor is left attached to the floor of fourth ventricle.
    E. If extending through foramen of Luschka, ipsilateral tonsil is elevated to expose tumor in cerebello-medullary cistern.

✅ **Answer B**
    — Ependymoma is quite soft and does not invade the pia of surrounding cerebellum or spinal cord and does not invade deep in brainstem at points of attachment caudal to the stria medullaris.

❓ **41. Surgery for ependymoma of the fourth ventricle.**
    **Positive prognostic factors. False answer is:**
    A. Smaller postoperative residual tumor.
    B. Female patient.

C. Laterally originating ependymoma.
D. No neuraxis dissemination.
E. Non-anaplastic tumors.

✅ **Answer C**
- Midline tumor originating from the floor of the fourth ventricle have better prognosis than laterally originating tumors arising near Luschka.

❓ **42. Surgery of exophytic cervicomedullary brainstem gliomas.**
**False answer is:**
A. Rarely extend above the pontomedullary junction.
B. Gross total excision can be easily achieved especially for medullary part of the tumor.
C. Postoperatively, most feared complication is respiratory difficulties.
D. Myelotomy is made in the middle of tumor as determined by intra-op ultrasound.
E. Any deviation in HR or BP might indicate manipulation near-critical medullary centers.

✅ **Answer B**
- Generally, the tumor should be carefully removed from the inside out. Resection till the tumor-brain interface is achievable in pontine or cervical lesions, but it is advised to stop after removing only 70–80% of medullary lesions to avoid violating normal medullary structures which might yield disastrous morbidities.

❓ **43. Surgery for choroid plexus papilloma of the fourth ventricle.**
**False answer is:**
A. More common in adults.
B. Pre-op angiography study (CTA or MRA) is essential to plan early devascularization.
C. PICA for tumors within cavity and occasionally AICA for tumors in lateral recess.
D. Gross total excision is possible as tumor typically does not invade cerebellum or floor.
E. Avulsion of the tumor from its pedicle can provide fast excision with minimal bleeding.

✅ **Answer E**
- Draining veins may be arterialized due to intratumoral shunting producing significant hemorrhage if torn.

**? 44. Surgery for choroid plexus carcinoma of the fourth ventricle.**
**False answer is:**
A. Occasionally involve the fourth ventricle.
B. Often occur during the first few days of life.
C. Degree of resection does not affect prognosis.
D. Shows wide CSF and neuraxis dissemination at or shortly after presentation.
E. Less homogeneous on neuroimaging than choroid papilloma due to necrosis, intratumoral hemorrhage, and cysts.

**✓ Answer C**
— Although generally choroid plexus carcinoma carries grave prognosis, 5-year survival rate after gross total resection is 86% in comparison to only 26% after partial resection.

**? 45. Surgery for dermoid cyst of the fourth ventricle.**
**False answer is:**
A. Dermal sinus is excised only if neuraxis connection is confirmed by radio imaging.
B. Total excision should be targeted since subtotal resection usually leads to recurrence.
C. Capsule violation spills cyst contents leading to severe chemical meningitis.
D. If rupture occurs, intra-op local steroid irrigation, and post-op systemic steroids are used.
E. If densely adherent to critical structures, bipolar coagulation of remnant capsule minimize recurrence.

**✓ Answer A**
— If a dermal sinus tract is present, it must be explored even if imaging does not indicate any abnormality.
— Absence of bony defect on exploration excludes possibility of intracranial extension.
— An elliptical incision is made around the opening of the dermal sinus tract and the tract is then removed en block.

**? 46. Surgery of hemangioblastoma of the fourth ventricle.**
**False answer is:**
A. In cystic tumors, cyst is fenestrated early to provide space for excising mural nodule.
B. Cyst wall is made up of compressed gliotic cerebellum rather than true tumor tissue.

C. Preoperative radiological studies of the arterial supply are imperative to identify feeders.
D. Solid tumors can be removed in a piece meal fashion.
E. Micro-Doppler evaluation of draining veins gives information about the amount of residual arterial supply and the progress in devascularization.

✅ **Answer D**
- A very difficult-to-control hemorrhage is expected if a solid tumor is penetrated before complete devascularization.
- These lesions must be deprived circumferentially from their arterial supply before interrupting their draining veins, and then removed in one piece.

❓ 47. **Removal of tumors adherent to the floor of fourth ventricle.**
   **False answer is:**
A. Total resection should not be attempted due to the risk of permanent cranial nerve deficits.
B. Areas of adhesion are disconnected from main tumor bulk using bipolar.
C. Later, the residual tumor tissue can be carefully thinned out by microsuction or Cavitron ultrasonic surgical aspirator.
D. Alternatively, sharp shaving of the residual tissue along with a superficial layer from the fourth ventricular floor can be attempted to attain gross total resection.
E. The thin remaining lining can be coagulated with low bipolar to reduce its viability.

✅ **Answer D**
- Sharp shaving of the residual tissue along with a superficial layer from the fourth ventricular floor should not be attempted to attain gross total resection. This is because the fourth ventricular floor contains critical brainstem structures that can affect the outcome if violated.

# CSF and Hydrocephalus

*Eleni Tsianaka, Sara A. Mohammad, Mustafa Ismail, and Awfa Aktham*


H. R. Salih et al. (eds.), *Cerebral Ventricles*, https://doi.org/10.1007/978-3-031-42342-0_13

**?  1.  Hydrocephalus, the FALSE is:**
    A. Consists of pathological CSF accumulation in the ventricular system of the brain.
    B. CSF overproduction is a rare etiological factor.
    C. Infection is the most common etiological factor for non-obstructive hydrocephalus.
    D. Non-obstructive hydrocephalus is mainly due to defective CSF overproduction.
    E. Constitutional ventriculomegaly does not need any treatment.

✓ **Answer D**
    — Non-obstructive hydrocephalus is due to defected CSF reabsorption from the arachnoid granulations.

**?  2.  Clinical manifestations of hydrocephalus, the FALSE answer is:**
    A. Slowly progressive hydrocephalus in adults can be asymptomatic.
    B. Chronic papilledema can cause blindness.
    C. Can include all the symptoms of increased ICP.
    D. Acute hydrocephalus can lead to brain herniation.
    E. Parinaud syndrome is a feature of well-tolerated chronic hydrocephalus.

✓ **Answer E**
    — Parinaud syndrome can develop at severe untreated cases of hydrocephalus.

**?  3.  Clinical manifestations of hydrocephalus in young children, the FALSE answer is:**
    A. Occipito-frontal circumference should be part of the routine follow-up of all children.
    B. Increased ICP can cause enlargement of scalp veins.
    C. Tendon reflexes are always normal.
    D. "Setting sun" sign is characteristic.
    E. Macewen's sign is positive.

✓ **Answer C**
    — Patient can have hyperreflexia.

**?  4.  Hydrocephalus and imaging, the FALSE answer is:**
    A. The ratio of frontal horns width divided by internal diameter should normally be less than 40%.
    B. The third ventricle is slit shape in normal individual.

C. Evans ratio >0.3 is diagnostic for ventriculomegaly.
D. Erosions of sella turcica can be a feature of chronic hydrocephalus.
E. In chronic hydrocephalus, the third ventricle is normal.

✅ **Answer E**
- In chronic hydrocephalus, the third ventricle is herniating into the sella turcica.

❓ 5. **Differential diagnosis of hydrocephalus, the FALSE answer is:**
A. Hydrocephalus ex vacuo is due to cerebral atrophy.
B. Hydranencephaly is more often due to bilateral ICA infarcts.
C. Hydrocephalus ex vacuo does not need surgical treatment.
D. Hydranencephaly after shunting can be improved, due to re-expansion of the brain cortex.
E. Hydrocephalus ex vacuo represents normal CSF hydrodynamics.

✅ **Answer D**
- Hydranencephaly must be differentiated from the severe hydrocephalus, which can be improved after shunting, because of brain cortex re-expansion.

❓ 6. **Benign external hydrocephalus, the FALSE answer is:**
A. There is an enlargement of the subarachnoid spaces in the first year of life.
B. The head circumference can be enlarged.
C. Cortical vein sign is used for differential diagnosis with subdural hematoma.
D. Brain ventricles are significantly enlarged.
E. Usually does not need a shunt.

✅ **Answer D**
- Brain ventricles are mildly enlarged or normal in size.

❓ 7. **Idiopathic normal pressure hydrocephalus (NPH), the FALSE answer is:**
A. It is the most common cause of hydrocephalus in adults.
B. "Freezing" of gait can be a clinical feature.
C. Dementia is of subcortical type.
D. On the MRI, callosal angle is <40°.
E. Sylvian fissures are dilated.

✅ **Answer D**
- On MRI, callosal angle of the patients with idiopathic NPH varies between 40° and 90°.

**8. Idiopathic NPH prognosis, the FALSE answer is:**

A. Dementia development suggests unfavorable outcome.
B. Cerebral atrophy suggests unfavorable outcome.
C. CSF outflow resistance >18 mmHg/mL per minute suggests unfavorable outcome.
D. Multiple white matter lesions suggest unfavorable outcome.
E. Early diagnosis and treatment suggest favorable outcome.

**Answer C**

- CSF outflow resistance >18 mmHg/mL per minute suggests favorable outcome.

**9. Infratentorial tumors and hydrocephalus, the FALSE answer is:**

A. Infratentorial tumors often present with the clinical picture of hydrocephalus.
B. Shunting before the main surgery can lead to "shunt metastasis."
C. Excessive rapid CSF drainage can lead to upward transtentorial herniation.
D. Ventriculostomy at the time of the main surgery must be avoided.
E. Upward transtentorial herniation can be avoided by not proceeding with CSF drain before dura opening.

**Answer D**

- Ventriculostomy at the time of the main surgery can be an option.

**10. Hydrocephalus and pregnancy, the FALSE answer is:**

A. Hydrocephalus is not a contraindication for pregnancy.
B. Conservative treatment includes bed rest and fluids restriction.
C. Steroids and diuretics are absolute contraindication.
D. During the third trimester, ventriculo-peritoneal shunt must be avoided.
E. In asymptomatic patient, vaginal delivery is preferable.

**Answer C**

- Steroid and diuretics can be used in cases of severe symptoms of increased ICP.

**11. The fourth ventricle, the FALSE is:**

A. Tent-shaped midline cavity located between the cerebellum and the brainstem.
B. It is dorsal to pons and medial to cerebellar peduncles.

    C. It is connected through the aqueduct with the third ventricle.
    D. It is connected through foramen of Magendie with cisterna magna.
    E. Most of the cranial nerves arise near its roof.

✓ **Answer E**

— Most of the cranial nerves arise near its floor. It has a roof, a floor, and two lateral recesses.

**? 12. The fourth ventricle, the FALSE is:**
    A. Roof expands laterally and posteriorly just below the aqueduct.
    B. At lateral recess level, it has the greatest height and width.
    C. Apex of its roof is called the fastigium.
    D. Inferior fastigium is thicker than superior part.
    E. Superior part of fastigium is formed by neural structures.

✓ **Answer D**

— The superior part is distinctly different from the inferior part, in that the inferior part is formed largely by thin membranous layers and the superior part is formed by thicker neural structures.

**? 13. The roof of fourth ventricle, the FALSE is:**
    A. Its superior part is divided into two lateral and single median part.
    B. Median parts are formed by the superior medullary velum.
    C. Lateral parts are formed by inner surface of cerebellar peduncles.
    D. Superior medullary velum is continuous at fastigium with inferior medullary velum.
    E. Caudal portion of each lateral wall is formed by superior cerebellar peduncle.

✓ **Answer E**

— The rostral portion of the ventricular surface of each lateral wall is formed by the medial surface of the superior cerebellar peduncle, and the caudal part is formed by the inferior cerebellar peduncle.

**? 14. Cerebellar peduncle and fourth ventricle; the FALSE is:**
    A. Middle peduncle is separated from ventricular surface by inferior and superior peduncles.
    B. Fibers of inferior peduncle line ventricular surface of superior margin of lateral recess.

C. Fibers of inferior cerebellar peduncle ascend in the posterolateral midbrain.
D. Fibers of superior peduncle arise in dentate nucleus.
E. Fibers of superior peduncle form ventricular surface of superior part of lateral wall.

✓ **Answer C**
 - The fibers of the inferior cerebellar peduncle ascend in the posterolateral medulla and turn posteriorly in the inferomedial part of the fiber bundle formed by the union of the three peduncles.

? **15. Upper part of fourth ventricle's roof, the FALSE is:**
A. Cisternal surface of the roof form posterior wall of the cerebellopontine fissure.
B. Cerebellomesencephalic fissure is also called as precentral cerebellar fissure.
C. Lingula is a thin narrow tongue of vermis sits on surface of superior medullary velum.
D. Interpeduncular sulcus marks junction of superior and middle cerebellar peduncles.
E. Trochlear nerves arise in cerebellomesencephalic fissure below inferior colliculi.

✓ **Answer A**
 - The cisternal (external) surface of the structures forming the superior part of the roof and also form the anterior wall of the cerebellomesencephalic fissure.

**13**

# CSF Diversion

*Laith T. Al-Ameri, Noor F. Hassan, Ali Q. Alforati, and Mustafa A. Shamkhi*

**? 1. VP shunt complications, the FALSE answer is:**
A.  Shunt failure is the highest in the first 6 months following insertion.
B.  Infection is the most common.
C.  Pseudocyst is a late complication.
D.  Subdural hematomata occur in over shunting for low pressure hydrocephalus.
E.  Bowel perforation is a rare complication.

**✓ Answer B**
— Obstruction is the most common type of complication followed by infection.

**? 2. Infection as a complication of VP shunt, the FALSE answer is:**
A.  Reduced significantly by antibiotic-impregnated shunt catheters.
B.  Contamination of the shunt system at the time of surgery is the main cause.
C.  *Staphylococcus aureus* is the main causative organism.
D.  Previous infection is a risk factor.
E.  Double gloving during surgery is associated with less infection.

**✓ Answer C**
— *Staphylococcus epidermidis* and coagulase negative staphylococci are the most common organisms.

**? 3. ICH related to shunt surgery, the FALSE answer is:**
A.  8% of patients show ICH by CT scan postoperatively.
B.  Only 2% of patients show symptomatic ICH.
C.  EVD shows the lower rate of ICH.
D.  Elective VP shunt placement shows the lower rate of ICH.
E.  Misplaced ventricular catheter associated with higher rate ICH.

**✓ Answer C**
— EVD shows the higher rate of ICH.

**? 4. Clinical predictors of VP shunt block in pediatric age group include, the FALSE answer is:**
A.  Drowsiness.
B.  Altered mental status.
C.  Nausea and vomiting.
D.  Fever.
E.  Increased head circumference.

### ✅ Answer D

- Fever is less reported as a predictor of VP shunt block, it may suggest an alternative diagnosis.

**? 5. Regarding VP shunt obstruction, the FALSE answer is:**
   A. Is commonest VP shunt complication.
   B. Distal catheter is the most common site.
   C. High RBC in CSF shows no association to obstruction.
   D. Programmable valve shows risk reduction in proximal catheter obstruction only.
   E. Placement of proximal catheter in frontal horn is associated with lower malfunction rate.

### ✅ Answer B

- Proximal catheter is the most common site for obstruction. This may be due to proximal catheter clogged with brain parenchyma, and choroid plexus that present near the foramen of Monro, this explains the lower rate of malfunctions when placing the proximal catheter in the anterior horn away from choroid plexus.

**? 6. Rare complications of VP shunt, the FALSE answer is:**
   A. Bowel perforation.
   B. Calcification of shunt.
   C. Distal catheter migration into scrotum.
   D. Abdominal pseudocyst.
   E. Proximal catheter clogged with brain parenchyma.

### ✅ Answer E

- Obstruction is the most common complication with the proximal catheter block being the most common site for obstruction.

**? 7. The upper end ventricular shunt obstruction is high in the following locations, the FALSE answer is:**
   A. Septum pellucidum.
   B. Contacting the ipsilateral ventricle wall.
   C. In the third ventricle.
   D. Contacting the contralateral ventricular wall.
   E. Superior to the foramen of Monro.

### ✅ Answer E

- Locations of ventricular catheter tip superior to the foramen of Monro and at the body of lateral ventricle are associated with very low incidence of shunt malfunctions.

**8. Over-drainage in VP shunt, the FALSE answer is:**
- A. Subdural hematoma is a complication.
- B. Acute over-drainage increases the risk of slit ventricle syndrome.
- C. May be related to a distal catheter siphoning effect.
- D. Antisiphon valve device may reduce over-drainage complication.
- E. Lumbo-peritoneal shunt shows reduction of failure rate in patient with slit ventricle syndrome.

**Answer B**
- It is chronic over-drainage rather than acute that is thought to be a cause of slit ventricle syndrome.

**9. Factors that adversely impact the shunt survival time, the FALSE answer is:**
- A. Patients with previous external drainage system.
- B. Patients undergoing surgical excision of tumor.
- C. Type of hydrocephalus.
- D. Hydrocephalus following cranial surgery.
- E. Extreme age.

**Answer C**
- The type of hydrocephalus has no adverse impact on shunt survival time.

**10. Risk factors for ICH associated with EVD, the FALSE answer is:**
- A. Thrombocytopenia.
- B. INR greater than 1.4.
- C. Postoperative use of antiplatelet in first 24 h.
- D. Pre-placement antithrombotic use.
- E. Pediatric age group.

**Answer E**
- Age greater than 50 years identified as a risk factor rather than pediatric age group.

**11. Predictors of poor outcome for VP shunt in human immunodeficiency virus, the FALSE answer is:**
- A. Presence of immunosuppression.
- B. Cranial nerve palsies.
- C. Low preoperative GCS.
- D. Hypernatremia.
- E. Anemia.

✅ **Answer D**

- Severe hyponatremia is associated with poor outcome.

❓ **12. Lumbo-peritoneal shunt may be indicated in, the FALSE answer is:**
   A. NPH.
   B. Non-communicating hydrocephalus.
   C. Idiopathic intracranial hypertension.
   D. Spinal CSF leaks.
   E. Pseudo-meningoceles.

✅ **Answer B**

- The lumbo-peritoneal shunt is ineffective in non-communicating hydrocephalus management. While it is very effective in communicating type which is common following tubercular meningitis, chronic infections, and SAH.

❓ **13. Common complications of lumbo-peritoneal shunt include, the FALSE answer is:**
   A. Acquired Chiari malformation.
   B. CSF leaks.
   C. Over-drainage.
   D. Infection.
   E. Intracerebral hematoma.

✅ **Answer E**

- An intracerebral hematoma is a rare complication following lumbo-peritoneal shunt.

❓ **14. Risk factors for developing postoperative hydrocephalus in pediatric age group, the FALSE answer is:**
   A. MB.
   B. Lesions within the fourth ventricle.
   C. Brain stem compression.
   D. Astrocytoma.
   E. Age ≤ 2 years.

✅ **The answer is D.**

- Astrocytoma is a negative predictor for the development of hydrocephalus.

❓ **15. ETV is indicated in, the FALSE answer is:**
   A. NPH.
   B. Simplification of septated hydrocephalus.

    C. Colloid cyst removal.
    D. Hydrocephalus secondary to pineal tumors.
    E. Prematurity.

✓ **Answer E**
  − Prematurity is considered as a contraindication in ETV.

**? 16. Contraindications for ETV, the FALSE answer is:**
    A. Previous radiotherapy.
    B. Distorted ventricular anatomy.
    C. Myelomeningocele.
    D. IVH.
    E. Meningeal infection.

✓ **Answer C**
  − Although ETV for myelomeningocele is controversial, it is not contra-
    indicated.

**? 17. ETV, the FALSE answer is:**
    A. The confluence of thalamo-striate vein, septal vein, and choroid
       plexuses is a landmark of the foramen of Monro.
    B. Perforation of the third ventricular floor is performed between mam-
       millary bodies and infundibular recess.
    C. Fenestration should be made just posterior to the dorsum sellae.
    D. Sharp perforation of the floor is preferred.
    E. The Waterjet dissection technique is useful in the thick opaque floor.

✓ **Answer D**
  − Sharp perforation may lead to basilar artery injury. Blunt perforation
    is preferred to avoid the risk of vascular injury.

**? 18. MB in pediatrics, the FALSE answer is:**
    A. Obstructive hydrocephalus is a common presenting feature.
    B. Almost all patients will require permanent shunt.
    C. Patients with moderate to severe preoperative hydrocephalus may
       benefit from preoperative shunt.
    D. Upward trans-tentorial herniation may complicate preoperative ven-
       tricular drainage.
    E. EVD may be used preoperatively for hydrocephalus refractory to
       medical treatment.

✓ **Answer B**
  − Most patients will get their hydrocephalus resolved; a small propor-
    tion will need a permanent shunt.

**? 19. NPH, predictive test for shunt responsiveness (assessment of CSF hydrodynamics), the FALSE answer is:**
   A. Tap test is specific rather than sensitive.
   B. Urine incontinence is the symptom most likely to respond.
   C. External lumbar drainage is considered if tap test failed.
   D. Lumbar CSF infusion test is used for complex NPH.
   E. 10 mL/h is drained in lumbar CSF infusion test.

**✓ Answer B**
   — Among the three classic symptoms of NPH (impaired gait, urine incontinence, and dementia); impaired gait is the first symptoms to respond. Gait should be evaluated immediately before and 2–4 h following tapping.

**? 20. NPH, the FALSE answer is:**
   A. Is the most common type of hydrocephalus in adult.
   B. Incontinence without awareness is characteristic.
   C. CSF shunt is the only effective treatment.
   D. Programmable shunts reduce subdural hematoma significantly.
   E. Programmable shunts decrease the need for shunt revision.

**✓ Answer B**
   — In idiopathic NPH, patients are usually aware for their incontinence and urinary urge.

**? 21. Ventriculo-atrial shunt complications, the FALSE answer is:**
   A. More shunt obstructions than VP shunt.
   B. Excessive drainage.
   C. Autoimmune glomerulonephritis.
   D. Pulmonary embolism.
   E. Thrombosis of the superior vena cava.

**✓ Answer A**
   — The ventriculo-atrial shunt shows similar obstruction rate of upper catheter obstructions as VP shunt.

**? 22. Temporary management of acute hydrocephalus, the FALSE answer is:**
   A. Extra-ventricular drainage.
   B. Repeated lumbar drainage.
   C. Ventriculo-gallbladder shunt.
   D. Ventriculo-subgaleal shunt.
   E. Ommaya reservoir insertion.

✅ **Answer C**

- Ventriculo-gallbladder shunt is a permanent method of management used as an alternative shunt if VP shunt is contraindicated.

❓ **23. Ommaya ventricular reservoir, the FALSE answer is:**
   A. Used for temporary CSF tapping.
   B. Catheter should be placed in third ventricle.
   C. Indicated in preterm infants with IVH.
   D. Low risk of reservoir infection.
   E. Reduce risk of future shunt revisions.

✅ **Answer B**

- Catheter should be placed in lateral ventricle.
- Ommaya reservoir is used also for intracerebral injection such as chemotherapeutic agent and morphine as a palliative therapy.

❓ **24. Ventriculo-subgaleal shunt, the FALSE answer is:**
   A. Can be useful for preterm infants.
   B. Incision is performed near-anterior fontanelle.
   C. The distal end is between the galea and the periosteum.
   D. No valve is used.
   E. Is a permanent method for CSF diversion.

✅ **Answer E**

- Ventriculo-subgaleal shunt is a temporary shunt procedure for repeated CSF tapping until the preterm infant is fit for a permanent VP shunt.

❓ **25. Ventriculo-gallbladder shunt, the FALSE answer is:**
   A. Cholecystitis is a contraindication.
   B. Permanent method of CSF diversion.
   C. Bile duct diseases is a contraindication.
   D. Is the most common alternative to VP shunt.
   E. *Staphylococcus epidermidis* is the most common bacteria found as a complication.

✅ **Answer D**

- Ventriculo-gallbladder shunt is rarely used. Ventriculo-atrial is the most common alternative.

**? 26. Slit ventricle syndrome complicating shunt placement, the FALSE answer is:**
A. More in the elderly population.
B. Medium pressure valve shows more over-drainage than delta valve.
C. ETV is effective in management.
D. Characterized by slow refilling of the shunt reservoir.
E. Lumbo-peritoneal shunt is effective in management.

**✓ Answer A**
- Older patients show less incidence of slit ventricle syndrome due to decreased brain elasticity.

# Cerebral Ventricle: Congenital Lesions

Serge E. Mba, Fatimah O. Ahmed,
Ahmed I. Abdulwahhab, Rania H. Al-Taie,
Zahraa A. Alsubaihawi, and Mohammed Al-Dhahir

© The Author(s), under exclusive license to Springer Nature Switzerland AG 2023
H. R. Salih et al. (eds.), *Cerebral Ventricles*, https://doi.org/10.1007/978-3-031-42342-0_15

**? 1.  Holoproencephaly, the false answer is:**
   A.  Alobar holoproencephaly results from merger of lateral ventricles.
   B.  SHH mutation is responsible for most of the abnormalities.
   C.  It is present in 1 in 15,000 live births.
   D.  Alcohol abuse has not been implicated as a cause.
   E.  It is present in 5% of patients with Smith-Lemli-Opitz syndrome.

**✓ Answer D**
   – Alcohol abuse is thought to selectively destroy midline cells at the early stage of development.

**? 2.  Coarctation, the false answer is:**
   A.  Refers to apposition of two ventricular walls.
   B.  Focal coarctation can isolate a connatal cyst.
   C.  The incidence is less than 1%.
   D.  It results from inflammation or gliosis.
   E.  It occurs more commonly in the occipital horn of the lateral ventricle.

**✓ Answer D**
   – Coarctation is an error of development and does not result from inflammation nor gliosis.

**? 3.  Coarctation, the false answer is:**
   A.  Often requires active treatment.
   B.  Resolves spontaneously by 2 months of age.
   C.  Connatal cysts occur anterior to the foramen of Monro.
   D.  Can be differentiated from subependymal cysts.
   E.  Are benign lesions.

**✓ Answer A**
   – Ventricular coarctation often resolves spontaneously and does not require active treatment.

**? 4.  L1 Syndrome, the false answer is:**
   A.  It is inherited in an autosomal dominant manner.
   B.  Bilateral absence of the pyramids on imaging is pathognomonic.
   C.  Aqueductal stenosis may be a feature.
   D.  Patients often present with adducted thumb.
   E.  Sons are affected while daughters are carriers.

✅ **Answer A**

- L1 Syndrome is inherited in an X-linked manner.

❓ **5. X-linked hydrocephalus, radiographic features, the false answer is:**
   A. Symmetric hydrocephalus with posterior horn dilation.
   B. Large massa intermedia.
   C. Hypoplastic cerebellar vermis.
   D. Large quadrigeminal plate.
   E. Rippled ventricular wall seen before VP shunt insertion.

✅ **Answer E**

- The rippled ventricular wall follows ventricular peritoneal shunt insertion and is thought to be pathognomonic for X-linked syndrome.

❓ **6. X-linked hydrocephalus: management, the false answer is:**
   A. Ventricular peritoneal shunt improves neurologic outcome.
   B. There is no genetic therapy for L1CAM.
   C. Medical termination of pregnancy may be a consideration.
   D. Genetic testing is indicated in suspected patients.
   E. Management is symptomatic.

✅ **Answer A**

- Ventricular peritoneal shunt does not improve neurologic outcome but reduces head size and helps improve care of the baby by the caregiver.

❓ **7. Choroid plexus cysts, the false answer is:**
   A. Are usually bilateral and demonstrate signal characteristics similar to CSF.
   B. Should be differentiated from choroidal metastasis in adults.
   C. May be acquired or congenital.
   D. Most are found incidentally.
   E. Are usually benign.

✅ **Answer A**

- Choroid plexus cysts are usually unilateral with signal different from CSF, an incidental finding that is often linked to trisomy 18.

❓ **8. Aqueductal stenosis, the false answer is:**
   A. Stenosis often occurs in the caudal one-third.
   B. The adytum narrows to a dorsally based triangle shape.
   C. The adytum is deformed into a round opening when the fourth ventricle is trapped.

   D. A single ependymal layer lines the lumen of the aqueduct.
   E. Results in raised ICP.

✔ **Answer B**
- The adytum in the normal aqueduct has a characteristic dorsally based triangle that focally narrows to a funnel-like structure in stenosis.

❓ **9. Aqueductal stenosis, associated ventricular deformities, the false answer is:**
   A. Ventricular dilation.
   B. Focal enlargement of the third ventricle.
   C. Ventricular diverticula.
   D. Spontaneous ventriculo-cisternostomies.
   E. Periventricular cysts.

✔ **Answer E**
- Subependymal cysts occurs as a result of separation of the ependymal from the subependymal layer.

❓ **10. Aqueductal stenosis, radiology features, the false answer is:**
   A. Enlarged inferior recess of third ventricle.
   B. Abnormally thinned corpus callosum.
   C. Fourth ventricle diverticulum.
   D. Lateral ventricle diverticulum.
   E. Ventriculomegaly in the setting of rhomboencephalosynapsis.

✔ **Answer C**
- The diverticulum is found in the lateral ventricle.

❓ **11. Aqueductal stenosis, histopathologic classification, the false answer is:**
   A. Stenosis.
   B. Not otherwise specified.
   C. Forking.
   D. Septum formation.
   E. Gliosis.

✔ **Answer B**
- NOS does not feature in the classification of aqueductal stenosis.

❓ **12. Dandy Walker Malformation: Definition, the false answer is:**
   A. It is a congenital malformation.
   B. There is enlargement of the posterior fossa.
   C. There is enlargement of the fourth ventricle.
   D. The Dandy Walker variant has a normal size posterior fossa.
   E. It can be an acquired disorder.

**15**

✅ **Answer E**
- Dandy Walker Malformation and its variants are congenital malformations that the represent a continuum of developmental anomalies that are grouped together as Dandy-Walker complex.

❓ **13. Dandy-Walker Malformation: Epidemiology, the false answer is:**
  A. It is the most common cerebellar malformation.
  B. DWM is more common in males.
  C. It can be inherited as an autosomal recessive trait.
  D. Gestational exposure to TORCH infections is a risk factor.
  E. Incidence is 1 per 25.000 live birth.

✅ **Answer B**
- DWM is more common in female by a ratio of 3:1.

❓ **14. Dandy-Walker Malformation: Pathology, the false answer is:**
  A. Hypoplastic fourth ventricle.
  B. Hypoplastic vermis.
  C. Some association with FOXC1 gene.
  D. Atresia of the foramen of Lushka.
  E. Enlarged fourth ventricle.

✅ **Answer A**
- An enlarged fourth ventricle is the definition of DWM.

❓ **15. Dandy-Walker Malformation: epidemiology, the false answer is:**
  A. Prevalence is about 1 in every 30,000 births.
  B. It can be diagnosed in adulthood.
  C. Maternal use of warfarin is a risk factor.
  D. It is three times more common in females.
  E. There is no geographical distribution.

✅ **Answer B**
- DWM is a congenital malformation that is present at birth, the ability to have a fetal diagnosis allows the determination of whether the condition is DWM or a simple hydrocephalus due to trapped fourth ventricle.

❓ **16. Dandy-Walker Malformation, Associations, the false answer is:**
  A. Hydrocephalus in 80–90% of patients.
  B. Posterior fossa anomalies.
  C. Encephalocele in 20–70% of cases.
  D. PHACES syndrome.
  E. Myelomeningocele.

✅ **Answer E**

- Myelomeningocele is commonly associated with an Arnold Chiari type II malformation, not with DWS.

❓ **17. Dandy-Walker Syndrome: Evaluation, the false answer is:**
   A. Neuropsychological tests are not required.
   B. Contrast MRI is the test of choice.
   C. US is a useful initial investigation.
   D. Vermian abnormalities are characteristic.
   E. Physical examination may reveal neurocutaneous melanosis.

✅ **Answer A**

- All patients should be assessed for their cognitive level of functioning to direct further care. 8–10% of patients with DWM have neurocutaneous melanosis.

❓ **18. Congenital causes of hydrocephalus, the false answer is:**
   A. Aqueductal stenosis.
   B. Type 1 Chiari malformation.
   C. Dandy-Walker cyst.
   D. Platybasia.
   E. Germinal matrix hemorrhage.

✅ **Answer E**

- Germinal matrix hemorrhage occurs after delivery in premature babies.

❓ **19. In aqueductal stenosis, the false answer is:**
   A. The posterior fossa is small.
   B. Hydrocephalus can be manifest in utero.
   C. Bulging fontanelle is not a feature.
   D. MacEwan's crackpot sign can be elicited.
   E. Setting sun eyes is a feature.

✅ **Answer C**

- Bulging fontanelle is a feature of hydrocephalus and is a prominent feature in aqueductal stenosis.

❓ **20. Arachnoid cyst, the false answer is:**
   A. Can cause hydrocephalus by obstruction of CSF in the third ventricle.
   B. Simple cyst that does not secrete CSF.
   C. Comprise 1% of intracranial masses.

D. Suprasellar cyst with hydrocephalus can present with macrocephaly.

E. Shunting of cyst is probably the best overall treatment.

✓ **Answer B**

— Histologically, simple cysts are lined with cells capable of secreting CSF.

**21. Hydranencephaly, the false answer is:**

A. Is a neurulation defect.

B. Is a post neurulation defect.

C. Is rarely associated with facial dysmorphism.

D. Most commonly due to bilateral ICA infarcts.

E. May mimic maximal hydrocephalus.

✓ **Answer A**

— Hydranencephaly is a post neurulation defect and thought to be due to bilateral ICA infarcts.

**22. Hydranencephaly differs from hydrocephalus, the false answer is:**

A. Hydrocephalus may respond to shunting.

B. EEG shows no cortical activity in maximal hydrocephalus.

C. Frontal lobe and frontal horns are not visible on CT in hydranencephaly.

D. Flow void in supraclinoid carotid on angiogram may be seen in hydranencephaly.

E. Transillumination is not a useful differentiating factor.

✓ **Answer B**

— EEG is one of the best ways to differentiate the two conditions, it shows electrical activity in hydrocephalus but not in hydranencephaly.

**23. Holoprosencephaly, the false answer is:**

A. Is caused by failure of cleavage of telencephalon.

B. Associated with trisomy 13 in 80% of cases.

C. Most children do not survive beyond infancy.

D. The risk is increased in subsequent pregnancies.

E. Facio-cerebral dysplasia is uncommon.

✓ **Answer E**

— Also called arhinencephaly due failure of cleavage of telencephalon vesicle, the degree of craniocerebral dysplasia mirrors the severity of the holoprosencephaly, it is a common feature.

**? 24. Chiari 1. Investigations, the false answer is:**
A. Craniocervical junction abnormalities can be seen on plain X-ray.
B. Non-contrast CT scan is a useful first investigation.
C. MRI of brain and C-Spine is the test of choice.
D. Cine MRI may demonstrate CSF blockage at the foramen magnum.
E. Myelography coupled with CT shows good reliability.

**✓ Answer B**
— CT scan is poor at evaluating the craniocervical junction due to bony artifact, also in young children there is a need to limit radiation exposure.

**? 25. Aqueductal stenosis, etiology, the false answer is:**
A. Point mutation of the Xq28 locus (L1CAM gene).
B. Intrinsic pathology.
C. Unknown.
D. Use of warfarin.
E. Infection.

**✓ Answer D**
— Warfarin has been implicated in the development of Dandy-Walker Malformation but not directly linked to aqueductal stenosis.

**? 26. Dandy-Walker Malformation: Symptoms, the false answer is:**
A. Seizures in 15%.
B. Cognitive impairment in 50–70%.
C. Decreased ICP because of increased posterior fossa size.
D. Impaired motor function.
E. Macrocephaly in 80%.

**✓ Answer C**
— Raised ICP is frequent as hydrocephalus is associated with the condition in 70–90% of cases.

**? 27. Chiari 1, management, the false answer is:**
A. Early surgery is recommended.
B. Asymptomatic patients should be operated.
C. Cerebellar symptoms improve after surgery in 87% of patients.
D. Weakness symptoms do not respond well to surgery.
E. C1-C3 laminectomy is often required.

✅ **Answer B**

- Asymptomatic patients can be followed up and surgery should be offered only when a patient develops symptoms attributable to the malformation.

❓ **28. Chiari 1, associations, the false answer is:**
   A. Syringomyelia in 3%.
   B. Hydrocephalus in 9%.
   C. Craniosynostosis.
   D. Klippel-Feil syndrome.
   E. Basilar invagination.

❓ **Answer A**

- Syringomyelia is a prominent feature of Chiari 1 and is present in 30–70% of cases.

❓ **29. Chiari 1, epidemiology, the false answer is:**
   A. Females more than males.
   B. Average age at presentation is fourth decade.
   C. Symptoms last between 1 month and 20 years.
   D. Spontaneous improvement usually occur.
   E. Achondroplasia is an associated condition.

✅ **Answer D**

- Even though natural history is still a matter of debate, patient may remain static for years, rarely do they improve spontaneously.

❓ **30. Chiari 1, Signs, the false answer is:**
   A. Upbeat nystagmus.
   B. Foramen magnum syndrome in 22%.
   C. Central cord syndrome in 65%.
   D. Cerebellar signs in 11%.
   E. Normal neurology in 10%.

❓ **Answer A**

- Downbeat nystagmus is characteristic of Chiari 1, it indicates a structural lesion in the posterior fossa especially at the cervicomedullary junction.

❓ **31. Chiari 1, the false answer is:**
   A. It is also called primary cerebellar ectopia.
   B. Disruption of CSF flow at the foramen magnum.
   C. It is always a congenital lesion.
   D. Displacement of the medulla is unusual.
   E. Hydrocephalus is rarely present.

✅ **Answer C**

- Chiari 1 can be congenital or acquired, any cause of raised ICP or lumbar decompression can precipitate a Chiari 1 malformation.

❓ **32. Dandy-Walker Malformation, treatment, the false answer is:**
  A. Shunting of posterior fossa.
  B. Concomitant shunting of lateral ventricle.
  C. ETV when aqueduct is patent.
  D. ETV when aqueductal stenosis.
  E. Observation for asymptomatic patients.

✅ **Answer D**

- ETV is not recommended in the presence of aqueductal stenosis.

❓ **33. Schizencephaly, the false answer is:**
  A. Unlike porencephalic cyst, it does not communicate with the ventricle.
  B. Typically, it is lined with cortical grey matter.
  C. It always communicates with the ventricle.
  D. Septum pellucidum is absent in 80–90% of cases.
  E. Seizure is a common presenting symptom.

✅ **Answer A**

- Both porencephalic and schizencephaly may communicate with the ventricle, schizencephaly always, porencephaly sometimes.

**15**

# Cerebral Ventricle: Infection

*Muthanna N. Abdulqader, Ruqaya A. Kassim, Hashim T. Hashim, Mustafa E. Almurayati, Nabeel H. Al-Fatlawi, and Mustafa Ismail*

**1. Blood–ependymal barrier, the false answer is:**
   A. *Escherichia coli* passes by transcellular passage.
   B. Protozoa pass by paracellular passage.
   C. *Mycobacterium tuberculosis* carried within a transmigrating leukocyte.
   D. Blood–ependymal barrier has ependymal cells replace blood–brain barrier (BBB) astrocyte foot processes.
   E. The commonest CNS-invasive pathogens to cross BBB are *Mycobacterium tuberculosis*.

**Answer E**
   – The most CNS-invasive pathogens to cross the BBB is *E. coli*.

**2. Ventriculitis, the false answer is:**
   A. Focal or diffuse inflammation of ependymal lining of the ventricle.
   B. Can't be distinguished from meningitis.
   C. More frequently in infants who undergo delayed closure of myelomeningocele.
   D. One of the causes of delayed neurological deterioration after aneurysmal SAH.
   E. Septations develop as an early sequela of ventriculitis.

**Answer E**
   – Septations develop as a late sequela.

**3. Pyogenic ventriculitis (ventriculitis secondary to meningitis), the false answer is:**
   A. More common in adults than infants.
   B. When meningitis fails to respond to antibiotics or recurs, ventriculitis should be considered.
   C. Typical organisms include Gram-negative species followed by Staph. species.
   D. Multiloculated hydrocephalus is more common with bacterial infections.
   E. Risk factors include low host immunity and higher virulence organism.

**Answer A**
   – More common in infants than adults.

**4. In bacterial ventriculitis CSF sampling, the false answer is:**
   A. Pleocytosis (over 10 cells/microL).
   B. Increase in CSF lactate.

    C. Positive culture or Gram stain.

    D. Low glucose (less than 25 mg/dL).

    E. Elevated protein count (greater than 50 mg/dL).

✅ **Answer C**

- Cultures may be negative following antibiotic therapy and in low virulence organisms like *Propionibacterium acnes*.

❓ 5. **Specific tests for (CSF, serum) in patient with ventriculitis, the false answer is:**

    A. Elevation in CSF lactate differentiates bacteria from aseptic meningitis.

    B. CSF lactate is superior to CSF glucose, protein, and leukocyte number.

    C. CSF lactate measurement is not affected by prior treatment with antibiotics.

    D. Elevated C-reactive protein (>20 mg/L) is useful in differentiating bacteria or viral etiology.

    E. Elevated *S. procalcitonin* differentiates bacterial infection from those surgeries.

✅ **Answer C**

- Antibiotics reduce the clinical value of CSF lactate measurement.

❓ 6. **Radiological finding of ventriculitis, the false answer is:**

    A. Non-contrast CT scan shows hyperdense ventricular debris seen in occipital horns.

    B. Periventricular low density represents reactive edema rather than ▶ trans-ependymal edema related to hydrocephalus.

    C. Ventricular debris on MRI is hypointense on T1 and hyperintense on T2.

    D. Restricted diffusion of ventricular debris seen on DWI/ADC.

    E. Thin uniform contrast enhancement of ependymal linings may be seen.

✅ **Answer C**

- Ventricular debris on MRI hyperintense on T1 and hypointense on T2.

❓ 7. **MRI finding in ventriculitis, the false answer is:**

    A. DWI and FLAIR are equally and highly sensitive for detecting debris and pus.

    B. Intraventricular debris and pus are the most frequent signs.

C. Periventricular hyperintensity is second most common contrast MRI feature.

D. Periventricular hyperintensity around ventricle horns are called "caps and rims."

E. Plexitis is seen as poorly defined margin and enhancing choroid plexus.

✅ **Answer C**

– Periventricular hyperintensity is best seen on FLAIR.

❓ 8. **Differential diagnosis of ependymal lining enhancement, the false answer is:**

(a) G▶ lioblastoma multiforme.

(b) Oligodendroglioma.

(c) ▶ Primary CNS lymphoma.

(d) Metastasis.

(e) ▶ Germinoma.

✅ **Answer B**

❓ 9. **Device-associated CNS infections, the false answer is:**

A. Two-third of healthcare-associated CNS infections are diagnosed within 2 weeks.

B. EVD and intraparenchymal ICP monitors have similar rate of infection.

C. Infection rates are generally similar between VP, VA, and VPL shunt.

D. Infection rates in ventricle devices is 8% in pediatric and 4–17% in adults.

E. Spinal cord medication pump infections are uncommon and occur in the first month.

✅ **Answer B**

– Intraparenchymal ICP monitors have a much lower risk of infection due to their structure (solid probe), tip location (white matter brain tissue), and reduced need for manipulation and handling after placement.

❓ 10. **Post-craniotomy (without EVD or shunt) ventriculo-meningitis, the false answer is:**

A. Antibiotic prophylaxis prevents post-craniotomy meningitis.

B. Meningitis is not acquired at the time of surgery but in postoperative period.

   C. 60–70% are caused by Gram-negative bacilli (*A. baumannii* and *E. coli*).

   D. Craniotomy is the most common surgery associated with postoperative Gram-negative meningitis.

   E. Avoid early re-interventions and CSF leak will prevent postoperative meningitis.

✅ **Answer A**

– Antibiotic prophylaxis does not prevent post-craniotomy meningitis.

❓ **11. Risk factor for deep wound infections after craniotomy, the false answer is:**

   A. Glasgow Coma Score of less than 10.

   B. Emergency surgery.

   C. CSF drainage and CSF leakage.

   D. Early reoperation.

   E. Shaving of the scalp limited to the incision.

✅ **Answer E**

– Total shaving of the scalp.

❓ **12. Intraventricular rupture of brain abscess, the false answer is:**

   A. Urgent craniotomy, debridement of abscess cavity and ventricular drainage.

   B. Abrupt worsening of preexisting headache with new onset meningismus.

   C. Layering debris in lateral ventricles and enhancement of the ependymal lining.

   D. Affects 15–25% of patients with brain abscess with a mortality rate of 35%.

   E. Hydrocephalus is found in 50% of cases.

✅ **Answer D**

– Intraventricular rupture of brain abscess affects 15–25% of patients with brain abscess with a mortality rate of 85%.

❓ **13. Risk factor of intraventricular rupture of brain abscess, the false answer is:**

   (a) Abscess size.

   (b) Multiloculated.

   (c) Decreased distance between the ventricle and the abscess.

   (d) Presence of localized ventricular enhancement on CT.

   (e) Deep seated.

✅ **Answer A**
- Neither the specific infecting organism nor abscess size is associated with intraventricular rupture of brain abscess.

❓ **14. Regarding shunt infection, the false answer is:**
A. The shunt infection rate is 4–8%.
B. Most shunt infections occur within the first 3 months.
C. *Staph. aureus* causes a slow indolent shunt infection and malfunction.
D. Bacteremia and shunt nephritis are established complications of VA shunt.
E. Low-virulence infections may not produce symptoms other than fever and leukocytosis.

✅ **Answer C**
- *Staphylococcus epidermidis* typically causes a slow indolent shunt infection and malfunction, whereas *Staphylococcus aureus* usually causes a more aggressive infection.

❓ **15. Laboratory study in shunt infection, the false answer is:**
A. CSF Gram stain and culture is gold standard for diagnosis of shunt infection.
B. Leukocytosis with predominance of (polymorphonuclear neutrophils) is typical.
C. Recently IVH or craniotomy increases CSF lymphocytes.
D. *Staph. aureus* decreases polymorphonuclear neutrophils and increases eosinophils.
E. Initial polymorphonuclear neutrophils predominance with later peak of lymphocytes.

✅ **Answer D**
- *P. acnes* decreases polymorphonuclear neutrophils and increases eosinophils.

❓ **16. Regarding the medical management of shunt infection, the false answer is:**
A. Initial empirical therapy includes vancomycin and ceftazidime.
B. The length of antibiotic use ranges from 10 to 14 days.
C. *Propionibacterium spp.* is treated with ampicillin or vancomycin.
D. Fungal shunt infection requires shunt removal and systemic amphotericin B.
E. Intravenous Garamycin can eliminate the biofilms of methicillin-resistant *Staphylococcus epidermidis* and methicillin-resistant *Staphylococcus aureus* (MRSA).

✅ **Answer E**
- I.V. Linezolid eliminates the biofilms of methicillin-resistant *Staphylococcus epidermidis* and MRSA.

❓ **17. Surgical treatment of shunt infection, the false answer is:**
   A. Medical management with antibiotics alone carries a high failure rate.
   B. Externalization of distal catheter and treatment with antibiotics employed in patients with abdominal symptoms.
   C. Complete removal of shunt and placement of EVD with antibiotic treatment is favored.
   D. Lumbar punctures or fontanelle taps can be performed for premature infants.
   E. Using the same entry site of previously infected shunt increases the risk of reinfection.

✅ **Answer E**
- Reusing the same entry site as the previously infected shunt catheter carries the same risk of reinfection of the shunt as did switching to a new entry site.

❓ **18. Catheter-related ventriculitis, the false answer is:**
   A. Associated with significant morbidity and mortality, especially with Gram-negative bacteria.
   B. *Staph. epidermidis* is the most common organism isolated 50–60%.
   C. The use of antibiotic targeting Gram-positive bacteria is a risk factor for the rise in Gram-negative bacteria.
   D. Following head trauma, *Staphylococcus aureus* is the most common pathogen.
   E. The incidence of catheter-related ventriculitis is 3–18%.

✅ **Answer D**
- Following head trauma, *Streptococcus pneumoniae* and Gram-negative rods are the most common pathogens.

❓ **19. Risk factors for catheter-related ventriculitis, the false answer is:**
   A. CSF leak.
   B. Neurosurgical operations.
   C. Frequent manipulation of the EVD system.
   D. Age.
   E. SAH.

✅ **Answer D**

- Age, gender, or race aren't risk factors.

❓ **20. EVD-related ventriculo-meningitis, the false answer is:**

A. Defined by a positive CSF culture with abnormal CSF findings or appropriate clinical signs and symptoms.
B. EVD contamination is positive CSF culture with abnormal CSF findings and lack of clinical signs other than fever.
C. Infection of skin and soft tissues at site of insertion is the most frequent complication.
D. Most common isolated Gram-negative species are enterobacteriaceae and pseudomonas.
E. *Candida ventriculitis* has increased due to broad spectrum antimicrobial therapy and compromised immune status.

✅ **Answer B**

- EVD catheter contamination is defined as positive CSF culture in the absence of abnormal CSF findings and lack of clinical signs other than fever.

❓ **21. Risk factor for EVD-related ventriculo-meningitis, the false answer is:**

A. EVD manipulation (e.g. CSF sampling, irrigation).
B. IVH.
C. EVD duration >11 days.
D. Surgical technique.
E. Drainage system leaks and disconnections.

✅ **Answer E**

❓ **22. Management of EVD-related ventriculitis, the false answer is:**

A. Antimicrobial therapy is continued for 10 to 14 days.
B. Initial therapy is an anti-staphylococcal agent with good CSF penetration.
C. For MRSA, doxycycline is highly recommended.
D. Pseudomonas and Acinetobacter are treated with cephalosporin or meropenem.
E. For fungal infections voriconazole, besides amphotericin B is a good option.

❓ **Answer C**

- For methicillin susceptible, therapy can be changed to flucloxacillin. For (MRSA or methicillin-resistant *Staphylococcus epidermidis*) vancomycin is highly recommended.

**?  23. Posttraumatic meningitis and ventriculitis, the false answer is:**
  A. The infection develops shortly after the injury or occurs months to years later.
  B. The most common cause of recurrent meningitis in children.
  C. The most common site for dural fistula is in the anterior cranial fossa.
  D. Surgery indicated when develop meningitis with persistent rhinorrhea.
  E. CSF rhinorrhea is distinguished from nasal secretions by Beta2-transferrin.

**✓ Answer B**
  – Traumatic head injury is the most common cause of recurrent meningitis in the adult while congenital fistulous connections are the most common causes in children.

**?  24. Intraventricular antibiotics, the false is:**
  A. Gentamycin 1–8 mg/day.
  B. Teicoplanin 5–40 mg/day.
  C. Vancomycin 5–20 mg/day.
  D. Ceftazidime 5–20 mg/day.
  E. Colistin 10 mg/day.

**✓ Answer D**

# Cerebral Ventricle: Vascular Anatomy

*Hayder R. Salih, Hatem Sadik, Sama Albairmani, Mustafa Ismail, and Samer S. Hoz*


H. R. Salih et al. (eds.), *Cerebral Ventricles*, https://doi.org/10.1007/978-3-031-42342-0_17

**?  1.   Arterial relation to the ventricles, the FALSE answer is:**
   A.  The ICA bifurcates below the frontal horn.
   B.  The apex of the basilar artery is situated between the temporal horns.
   C.  The anterior cerebral artery passes around the anterior wall of the third ventricle.
   D.  The posterior cerebral arteries pass below the temporal horns and atria.
   E.  The middle cerebral arteries pass below the frontal horns.

**✓ Answer D**
   – The posterior cerebral arteries pass medial to the temporal horns and atria and give rise to the posterior choroidal arteries, which pass through the choroidal fissure to supply the choroid plexus in the temporal horns, atria, and bodies.

**?  2.   Arterial relation to the ventricles, the FALSE answer is:**
   A.  All components of the circle of Willis are located below the frontal and body of the lateral ventricle.
   B.  The basilar artery apex locates below the floor of the third ventricle.
   C.  The anterior cerebral artery passes around the floor and anterior wall of the frontal horns.
   D.  The anterior cerebral artery sends branches into the floor of the lateral ventricle.
   E.  Posterior cerebral artery supplies atria and temporal horn choroid plexus.

**✓ Answer D**
   – Both the anterior and posterior cerebral arteries send branches into the roof. The middle cerebral arteries pass below the frontal horns to reach the sylvian fissures and then course over the insulae, where they are lateral to the bodies of the lateral ventricle.

**?  3.   The choroidal arteries, the FALSE answer is:**
   A.  They supply the choroid plexus of the lateral and third ventricles.
   B.  They arise from the internal carotid and posterior cerebral arteries.
   C.  They arise in the basal cisterns.
   D.  They reach the choroid plexus by passing through the choroidal fissures.
   E.  They have no supply to neural structures.

**✓ Answer E**
   – Each of the choroidal arteries gives off branches to the neural structures along its course.

**? 4. Choroidal artery, the FALSE answer is:**
   A. Lateral ventricle choroid plexus is supplied by both anterior and posterior choroidal artery.
   B. Every choroidal artery gives off branches to the neural structures along its course.
   C. Anterior choroidal arteries (AChA) supply a portion of the choroid plexus in atrium.
   D. Lateral posterior choroidal arteries supply the choroid plexus in the roof of the third ventricle.
   E. Medial posterior choroidal artery supplies part of the choroid plexus in the body of the lateral ventricle.

✓ **Answer D**
   ▬ The most common pattern for the choroidal arteries is as follows: The AChA supply a portion of the choroid plexus in the temporal horn and atrium. The lateral posterior choroidal arteries supply a portion of the choroid plexus in the atrium, body, and posterior part of the temporal horn. The medial posterior choroidal arteries supply the choroid plexus in the roof of the third ventricle and part of that in the body of the lateral ventricle.

**? 5. Choroidal arteries, the FALSE answer is:**
   A. AChA supply the third ventricle roof choroid plexus.
   B. AChA supply the body and temporal horn choroid plexus.
   C. Lateral posterior choroidal arteries supply the body and temporal horn choroid plexus.
   D. The lateral posterior choroidal artery supplies the atrium choroid plexus.
   E. The medial posterior choroidal artery supplies the body choroid plexus.

✓ **Answer A**
   ▬ See answer 4.

**? 6. Anterior choroidal artery (AChA), the FALSE answer is:**
   A. It arises from the ICA.
   B. It arises in the anterior incisural space.
   C. It courses laterally to reach the middle incisural space.
   D. It passes through the choroidal fissure near the inferior choroidal point.
   E. It passes dorsally along the plexus, reaching the foramen of Monro.

✅ **Answer C**

- The anterior choroidal artery arises from the ICA in the anterior incisural space and courses posteriorly to reach the middle incisural space.

❓ 7. **Lateral posterior choroidal artery, the FALSE answer is:**
   A. Arises in both ambient and quadrigeminal cisterns.
   B. Its branches enter the ventricle anterior to branches of AChA.
   C. It passes through the choroidal fissure at the level of fimbria, crus, and body of fornix.
   D. Arise from posterior cerebral artery or its cortical branches.
   E. Send branches to choroid plexus in the body of contralateral lateral ventricle.

✅ **Answer B**

- The lateral posterior choroidal artery's branches enter the ventricle behind the branches of the anterior choroidal artery.

❓ 8. **Medial posterior choroidal artery, the FALSE answer is:**
   A. Most frequently arises as 6 to 9 branches.
   B. Arises from posterior cerebral artery in the interpeduncular and crural cisterns.
   C. Courses in velum interpositum adjacent to internal cerebral veins.
   D. Supplies choroid plexus in the roof of the third ventricle.
   E. Supplies choroid plexus in the contralateral lateral ventricle.

✅ **Answer A**

- The medial posterior choroidal arteries most frequently arise as one to three branches.

❓ 9. **Deep venous drainage of the brain, the FALSE answer is:**
   A. Collects into channels in a subependymal location through walls of lateral and third ventricles.
   B. Veins from frontal horn, body, and surrounding gray and white matter drain into basal vein.
   C. Veins from temporal horn and adjacent periventricular structures drain into basal veins.
   D. Veins draining atrium and adjacent parts of the brain drain into the basal vein.
   E. Venous channels converge on the internal cerebral, basal, and great veins.

**17**

**✓ Answer B**

— The veins from the frontal horn, the body of the lateral ventricle, and the surrounding gray and white matter drain into the internal cerebral vein.

**? 10. Ventricular artery and veins, the FALSE answer is:**

A. In lateral ventricle surgery, veins provide orienting landmarks more than arteries.

B. On angiograms, arteries provide a less accurate estimate of the site and size of a lesion.

C. The ventricular veins are divided into medial and lateral groups.

D. Veins are more adherent to the ependymal and pial surfaces of the brain than arteries.

E. Arteries are larger and are easily visible through the ependyma.

**✓ Answer E**

— The veins are larger and are easily visible through the ependyma.

**? 11. The venous system of the ventricles, the FALSE answer is:**

A. The medial group of veins in the frontal horn consists of the anterior septal veins.

B. The lateral group in the frontal horn consists of the anterior septal vein, thalamocaudate, and posterior caudate veins.

C. The medial group of veins in the atrium and occipital horn consists of medial atrial veins.

D. The lateral group of veins in the atrium and occipital horn consists of lateral atrial veins.

E. The roof and lateral wall of the lateral ventricle is drained by transverse hippocampal veins.

**✓ Answer E**

— The floor of the lateral ventricle is drained by the transverse hippocampal veins while the roof and the lateral wall are drained predominantly by the inferior ventricular vein.

**? 12. Colloid cysts, the FALSE answer is:**

A. Most colloid cysts are supplied by branches of the anterior choroidal artery.

B. The internal cerebral vein has been commonly described as showing an anterior hump, with flattening and depression in its posterior two-thirds.

C. Hydrocephalus can cause outward bowing of the thalamostriate veins in the frontal venous phase.
D. Hydrocephalus can cause the increased sweep of the pericallosal artery in the lateral arterial phase.
E. Upward convexity in the initial segment of the internal cerebral vein is highly suggestive.

✅ **Answer A**
- Most colloid cysts are supplied by branches of the posterior medial choroidal artery.

❓ **13. Medial ventricular veins, the FALSE answer is:**
A. Ventricular veins drain into the internal cerebral, basal, and great veins.
B. The medial group consists of the anterior caudate vein in the frontal horn.
C. The lateral group includes thalamostriate, posterior caudate, and thalamocaudate veins in the body.
D. The lateral group includes atrial veins in the atrium and occipital horn.
E. Inferior ventricular and amygdalar veins are located in the temporal horn.

✅ **Answer B**
- The lateral group of the medial ventricular veins consists of the anterior caudate vein in the frontal horn.

❓ **14. Lateral ventricular veins, the FALSE answer is:**
A. Anterior septal veins are located in the frontal horn.
B. Posterior septal veins are located in the body.
C. Medial atrial veins are located in the atrium.
D. Transverse hippocampal veins are located in the occipital horn.
E. Transverse hippocampal veins drain into the anterior and posterior longitudinal hippocampal veins.

✅ **Answer D**
- Transverse hippocampal veins are located in the temporal horn.

❓ **15. Anterior septal vein, the FALSE answer is:**
A. Initially runs posteromedially on the anterior wall of the frontal horn and behind the genu of the corpus callosum.
B. At the anteromedial corner of the frontal horn, it runs backwards along the septum pellucidum and continues backwards along its lower border.

C. It follows the lateral border of the anterior column of the fornix.
D. It joins the internal cerebral vein at the anteroinferior margin of the foramen of Monro.
E. The anterior septal vein crosses the roof of the frontal horn and the body of the lateral ventricle, as well as the septum pellucidum.

✅ **Answer D**
 — The anterior septal vein joins the internal cerebral vein at the posterosuperior margin of the foramen of Monro.

❓ **16. Venous angle of the brain, the FALSE answer is:**
A. It is composed of the anastomosis of the thalamostriate and the anterior septal veins.
B. It is located along the posterior edge of the foramen of Monro.
C. False venous angle describes a TSV joining the ASV beyond the foramen of Monro.
D. It is also referred to as the "sharp angle."
E. At this point, the striothalamic vein turns laterally and superiorly.

✅ **Answer E**
 — At the venous angle, the striothalamic vein turns medially and slightly inferiorly.

❓ **17. Thalamostriate vein, the FALSE answer is:**
A. Receives several transverse caudate veins on its way toward the foramen of Monro.
B. Is the largest tributary of the internal cerebral vein.
C. Runs around the posterior tubercle of the thalamus at the level of the foramen of Monro.
D. Receives a vein from the head of the caudate nucleus.
E. Joins the internal cerebral vein at the posterosuperior margin of the foramen of Monro.

✅ **Answer C**
 — The thalamostriate vein runs around the anterior tubercle of the thalamus at the level of the foramen of Monro.

❓ **18. Basal veins of Rosenthal, the FALSE answer is:**
A. Are paired, paramedian veins which originate on the medial surface of the temporal lobe.
B. Pass lateral to the midbrain through the interpeduncular cistern to drain into the vein of Galen.
C. Are closely related to the posterior cerebral arteries (posterior cerebral arteries).

  D. Originate in the perimesencephalic cistern.
  E. Drain to the straight sinus and, subsequently, the confluence of sinuses.

✓ **Answer B**
 — The basal veins of Rosenthal Pass lateral to the midbrain through the ambient cistern to drain into the vein of Galen.

❓ **19. Internal cerebral veins, the FALSE answer is:**
  A. Are situated at the midline during the posterior transcallosal approach to the pineal recess.
  B. Upon reaching the pineal recess, they pass along the superolateral aspect of the pineal body.
  C. Unite superior to the splenium to form the vein of Galen.
  D. Are paired, paramedian veins that course posteriorly along the roof of the third ventricle.
  E. Usually, drain the thalami and periventricular white matter.

✓ **Answer C**
 — The internal cerebral veins unite inferior to the splenium to form the vein of Galen.

❓ **20. PICA, the FALSE answer is:**
  A. The first three branches are the predominant feeding vessels for fourth ventricular tumors.
  B. Five segments are: anterior medullary, lateral medullary, tonsillomedullary, telovelotonsillar, and cortical.
  C. Most of PICAs originate from the vertebral artery extradurally.
  D. The caudal loop is between the lower cranial nerves and the pole of the tonsil.
  E. The cranial loop is between the rostral pole of the tonsil and the inferior medullary velum.

❓ **Answer C**
 — Up to 20% of PICAs originate from the vertebral artery extradurally.

# Cerebral Ventricle: Vascular Lesions and Hemorrhage

*Osman Elamin, Ahmed Muthana, Hussein M. Hasan, Wamedh E. Matti, Hussein J. Kadhum, and Samer S. Hoz*

© The Author(s), under exclusive license to Springer Nature Switzerland AG 2023
H. R. Salih et al. (eds.), *Cerebral Ventricles*, https://doi.org/10.1007/978-3-031-42342-0_18

**18**

**? 1. Intraventricular hemorrhage (IVH), etiologies, the FALSE answer is:**
   A. Most occur because of the extension of intraparenchymal hemorrhages.
   B. Can result in a rupture of the aneurysm.
   C. SAH refluxing into the ventricles through foramina of Luschka or Magendie.
   D. In adults, it is an extension of subependymal hemorrhage.
   E. Can be associated with severe head trauma.

**✓ Answer D**
   – Extension of subependymal hemorrhage into ventricles is a common type of IVH in newborns.

**? 2. Pure or isolated IVH can result from, the FALSE answer is:**
   A. Ruptured AcommA aneurysm.
   B. Putaminal hemorrhages.
   C. Vertebral artery dissection.
   D. Intraventricular AVM.
   E. Intraventricular tumor.

**✓ Answer B**
   – IVH with putaminal hemorrhages is an example of the extension of intraparenchymal hemorrhages known as secondary IVH. Primary or pure IVH is confined only to the ventricles.

**? 3. IVH with ruptured aneurysm, the FALSE answer is:**
   A. Ruptured aneurysm accounts for ≈ 25% of IVH in adults.
   B. Carotid terminus aneurysms may rupture through the floor of the third ventricle.
   C. Carry the same prognosis of SAH without IVH.
   D. AcommA aneurysm can rupture through the lamina terminalis into the anterior third ventricle.
   E. Distal PICA aneurysms: may rupture directly into fourth ventricle through the foramen of Luschka.

**✓ Answer C**
   – Carry a worse prognosis (64% mortality).

**? 4. IVH, presentation, the FALSE answer is:**
   A. Sudden onset of severe headache.
   B. Nausea and vomiting.
   C. Decreased level of consciousness.
   D. Seizures are very common.
   E. Neurological deficit.

✅ **Answer D**
- Clinical seizures are uncommon but have been reported in IVH.

❓ **5. IVH, diagnosis, the FALSE answer is:**
A. MRI was performed immediately in suspected IVH.
B. MRI SWI is more sensitive for IVH than traditional GRE sequences.
C. CTA indicated in any patient <50–60 yrs.
D. MRI is more sensitive than CT to very small amounts of blood, especially in the posterior fossa.
E. Routine catheter angiography in the setting of primary IVH is warranted.

✅ **Answer A**
- Brain MRI is not highly sensitive in the first few hours while non-contrast CT is the mainstay of acute evaluation of patients suspected with IVH showed increased density ($\approx$ 30–80 HU).

❓ **6. IVH, general management, the FALSE answer is:**
A. ICU admission, elevating the head of bed, analgesia, mild sedation, MANNITOL.
B. Antiepileptic drug (AED) prophylaxis is controversial.
C. BP control target SBP < 110 mmHg.
D. Small hematomas and limited IVH usually do not need ICP treatment.
E. Early intensive lowering of BP does not result in a significant reduction of death or major disability but improves functional outcomes.

✅ **Answer C**
- Target SBP < 140 mmHg, keep MAP 70–110 mmHg.

❓ **7. IVH, surgical treatment, the FALSE answer is:**
A. Ventricular drainage as a treatment for hydrocephalus is reasonable, especially if decreased level of consciousness.
B. Endoscopic neurosurgical techniques for IVH evacuation may be advantageous compared with EVD.
C. CLEAR (Clot Lysis Evaluating Accelerated Resolution of IVH) III trial suggests that rtpa may be effective in reducing the volume of IVH.
D. Ventricular drainage is recommended in IVH associated with cerebellar hematoma.
E. Chronic hydrocephalus following IVH can be treated by VP shunt.

**18**

✅ **Answer D**

− EVD has a risk of upward herniation of cerebellum hemorrhage with IVH and does not relieve brainstem compression. Initial treatment with ventricular drainage rather the surgical evacuation is not recommended.

❓ 8. **IVH, prognosis, the FALSE answer is:**

    A. Patient age is considered as an important prognostic factor.
    B. Patients with lower Graeb scores were associated with better clinical conditions.
    C. A smaller volume of IVH is favorable.
    D. The development of hydrocephalus carries a poor prognosis.
    E. Secondary IVH showed a better prognosis than primary.

✅ **Answer E**

− The outcome of patients with primary IVH is better than those with secondary IVH.

❓ 9. **Intraventricular hemorrhage in newborn (IVH-n), alternate terms, the FALSE answer is:**

    A. Subependymal hemorrhage.
    B. Germinal matrix hemorrhage.
    C. Periventricular- IVH.
    D. Choroid plexus hemorrhage.
    E. Neonatal IVH.

✅ **Answer D**

− Choroid plexus hemorrhage can result in IVH, but not a term of IVH-n.

❓ 10. **IVH-n, pathogenesis, the FALSE answer is:**

    A. Birth asphyxia.
    B. Hypercapnia.
    C. Increased systematic venous pressure.
    D. Hypertension.
    E. Rapid i.v. resuscitation.

✅ **Answer D**

− Hypotension and hypoperfusion.

❓ 11. **IVH-n, anatomy of the germinal matrix, the FALSE answer is:**

    A. Located in the thick subependymal cell layer of the thalamostriate groove.
    B. Vulnerable watershed zone.

   C. Origin to both neural and glial cells.
   D. Progressive involution until 36 weeks.
   E. Supplied by terminal branches of posterior cerebral arteries.

✅ **Answer E**

– Supplied by Heubner's artery, terminal branches of the lateral striate arteries, and the anterior choroidal artery.

❓ **12. IVH-n, risk factors, the FALSE answer is:**
   A. Young gestational age.
   B. Increase CPP.
   C. Coagulopathy.
   D. Maternal cocaine use.
   E. Macrosomia.

✅ **Answer E**

– Low birth weight considers an important risk factor.

❓ **13. IVH-n, incidence, the FALSE answer is:**
   A. More in male.
   B. Low birth weight in 25%.
   C. Mortality rate 10%.
   D. Bimodal distribution.
   E. Mortality correlated to gestational age.

✅ **Answer C**

– Mortality can reach 55%.

❓ **14. IVH-n, prevention, the FALSE answer is:**
   A. Regular prenatal care.
   B. Avoid preterm labor.
   C. Avoid corticosteroid.
   D. Indomethacin.
   E. Surfactant.

✅ **Answer C**

– Antenatal corticosteroid administration used for prevention.

❓ **15. IVH-n, papile grading, the FALSE answer is:**
   A. Grade I: Hemorrhage restricted to subependymal region.
   B. Grade II: IVH without ventricular dilatation.
   C. Grade III: IVH with ventricular dilatation.
   D. Grade IV: IVH with parenchymal hemorrhage.
   E. Grade V: IVH with SAH.

**✓ Answer E**
- The grading system for primary IVH is four grades only.

**? 16. IVH-n, papile grading, the FALSE answer is:**
A. Grade I: seen in the 7 caudothalamic groove.
B. Grade II: poor prognosis.
C. Grade III: 20% mortality.
D. Grade IV: 36% mortality.
E. Grade IV: secondary to hemorrhagic infarction.

**✓ Answer B**
- Grade II overall good prognosis.

**? 17. IVH-n, presentation, the FALSE answer is:**
A. Flaccid paralysis.
B. Subclinical seizures.
C. Depressed fontanelle.
D. Hypotension.
E. Unreactive pupils.

**✓ Answer C**
- Tense fontanelle.

**? 18. IVH-n, hydrocephalus, the FALSE answer is:**
A. 50% of infants.
B. More in Grade I.
C. Young gestational age infants may be at higher risk.
D. Occurs 1 week after hemorrhage.
E. Most cases are due to cellular debris.

**✓ Answer B**
- Grades III and IV are more often associated with progressive ventricular dilatation than are lower grades.

**? 19. IVH-n, general measures, the FALSE answer is:**
A. Maintaining normal MAP.
B. Normalizing $pCO_2$.
C. Treating active hydrocephalus.
D. Daily LPs if needed.
E. Diuretic therapy.

**✓ Answer E**
- Furosemide and acetazolamide therapy are not safe and not effective in treating post-hemorrhagic ventricular dilatation.

**20. IVH-n, intervention includes, the FALSE answer is:**
   A. Lumbar drain.
   B. Lumbar puncture.
   C. Ventricular tap.
   D. Temporary ventricular access device.
   E. VP shunt.

✓ **Answer A**

**21. Intracranial aneurysms, associated IVH, the FALSE answer is:**
   A. The frontal horn hemorrhage and rupture of PcommA aneurysm.
   B. Third ventricle and rupture of AcommA aneurysm.
   C. Third ventricle and rupture of carotid terminus aneurysm.
   D. Fourth ventricle and rupture of PICA aneurysm.
   E. Lateral ventricle and rupture of AcommA aneurysm.

✓ **Answer A**
   – Ruptured PcommA aneurysm will cause carotid cistern, optic cistern, and sylvian cistern SAH; also, it can cause a subdural hematoma.

**22. Intraventricular vascular lesions includes the following, the FALSE answer is:**
   A. AVM.
   B. Cavernoma.
   C. Carotid-cavernous fistula.
   D. Telangiectasia.
   E. Venous angiomas.

✓ **Answer C**

**23. Intraventricular vascular lesions, characteristics, the FALSE answer is:**
   A. No sex predominance.
   B. Common in adults.
   C. Third ventricle most common site.
   D. Mimic tumor on radiology.
   E. Associated with congenital anomalies.

**The answer is C.**
   – The lateral ventricle is the most frequent origin, followed by the third ventricle.

**② 24. Intraventricular aneurysms, characteristics, the FALSE answer is:**
A. Presented with isolated IVH.
B. Giant aneurysms.
C. The first option for management is endovascular.
D. Very rare.
E. Good prognosis.

**✓ Answer B**
- Intraventricular aneurysms are frequently very small (<5 mm diameter).

**② 25. Intraventricular Cerebral Cavernomas (IVCs), characteristics, the FALSE answer is**
A. Occur in Only 2–10% of Patients with Cerebral Cavernomas
B. It Is Genetically Inherited Disorder
C. Good prognosis.
D. Equally in Males and Females
E. Almost all Patients Presented with an Acute Headache on Admission

**✓ Answer B**
- Reports concerning IVC are scarce and are limited mostly to sporadic case reports.

**② 26. (IVCs), characteristics, the FALSE answer is**
A. The Mean Age of the Patients Is 36.5 Years
B. Can Be Presented by IVH
C. The most Frequent Location Is the Fourth Ventricle
D. Complete Surgical Resection Is the Treatment of Choice
E. The Microsurgical Approach Is Currently Considered the Gold Standard for IVC Resection

**✓ Answer C**
- The most frequent location is the lateral ventricle about 52.6%.

# Supplementary Information

**References – 000**


H. R. Salih et al. (eds.), *Cerebral Ventricles*, https://doi.org/10.1007/978-3-031-42342-0

# References


1.  Tavighi SH, Saadatfar Z, Shojaei B, Rassouli MB. Brain ventricle development in *H. huso* (Beluga sturgeon) larvae. Anat Sci Int. 2016;91:350–7.
2.  Le Gars D, Lejeune JP, Peltier J. Surgical anatomy and surgical approaches to the lateral ventricles. In: Advances and technical standards in neurosurgery. Vienna: Springer; 2009. p. 147–87.
3.  Çırak M, Yağmurlu K, Kearns KN, Ribas EC, Urgun K, Shaffrey ME, Kalani MY. The caudate nucleus: its connections, surgical implications, and related complications. World Neurosurg. 2020;139:e428–38.
4.  Danaila L. Primary tumors of the lateral ventricles of the brain. Chirurgia (Bucur). 2013;108(5):616–30.
5.  Mark LP, Daniels DL, Naidich TP. The fornix. AJNR Am J Neuroradiol. 1993;14(6):1355.
6.  Senova S, Fomenko A, Gondard E, Lozano AM. Anatomy and function of the fornix in the context of its potential as a therapeutic target. J Neurol Neurosurg Psychiatry. 2020;91(5):547–59.
7.  Yeo SS, Jang SH. The effect of thalamic hemorrhage on the fornix. Int J Neurosci. 2011;121(7):379–83.
8.  Fogwe LA, Reddy V, Mesfin FB. Neuroanatomy, hippocampus. Treasure Island (FL): StatPearls Publishing; 2018.
9.  Mooshagian E. Anatomy of the corpus callosum reveals its function. J Neurosci. 2008;28(7):1535–6.
10.  Sarwar M. The septum pellucidum: normal and abnormal. Am J Neuroradiol. 1989;10(5):989–1005.
11.  Griffiths PD, Batty R, Reeves MJ, Connolly DJ. Imaging the corpus callosum, septum pellucidum and fornix in children: normal anatomy and variations of normality. Neuroradiology. 2009;51:337–45.
12.  Mithani K, Davison B, Meng Y, Lipsman N. The anterior limb of the internal capsule: anatomy, function, and dysfunction. Behav Brain Res. 2020;387:112588.
13.  Scelsi CL, Rahim TA, Morris JA, Kramer GJ, Gilbert BC, Forseen SE. The lateral ventricles: a detailed review of anatomy, development, and anatomic variations. Am J Neuroradiol. 2020;41(4):566–72.
14.  Sahana D, Rathore L, Kumar S, Sahu RK. Endoscopic anatomy of lateral and third ventricles: a must know for performing endoscopic third ventriculostomy. Neurol India. 2021;69(1):45.
15.  Kawashima M, Li X, Rhoton AL Jr, Ulm AJ, Oka H, Fujii K. Surgical approaches to the atrium of the lateral ventricle: microsurgical anatomy. Surg Neurol. 2006;65(5):436–45.
16.  Sánchez JJ, Rincon-Torroella J, Prats-Galino A, de Notaris M, Berenguer J, Rodríguez EF, Benet A. New endoscopic route to the temporal horn of the lateral ventricle: surgical simulation and morphometric assessment. J Neurosurg. 2014;121(3):751–9.
17.  Altafulla JJ, Suh S, Bordes S, Prickett J, Iwanaga J, Loukas M, Dumont AS, Tubbs RS. Choroidal fissure and choroidal fissure cysts: a comprehensive review. Anat Cell Biol. 2020;53(2):121–5.

18. Wen HT, Rhoton AL, de Oliveira E Jr. Transchoroidal approach to the third ventricle: an anatomic study of the choroidal fissure and its clinical application. Neurosurgery. 1998;42(6):1205–17.

19. Zemmoura I, Velut S, François P. The choroidal fissure: anatomy and surgical implications. In: Advances and technical standards in neurosurgery. Vienna: Springer; 2012. p. 97–113.

20. Mortazavi MM, Griessenauer CJ, Adeeb N, Deep A, Shahripour RB, Loukas M, Tubbs RI, Tubbs RS. The choroid plexus: a comprehensive review of its history, anatomy, function, histology, embryology, and surgical considerations. Childs Nerv Syst. 2014;30:205–14.

21. Castro FD, Reis F, Guerra JG. Lesões expansivas intraventriculares à ressonância magnética: ensaio iconográfico-parte 1. Radiol Bras. 2014;47:176–81.

22. Winn HR. Youmans and Winn neurological surgery. Philadelphia: Elsevier; 2017. p. 3525–9.

23. Barreto AS, Vassallo J, Queiroz LD. Papillomas and carcinomas of the choroid plexus: histological and immunohistochemical studies and comparison with normal fetal choroid plexus. Arq Neuropsiquiatr. 2004;62:600–7.

24. Wrede B, Hasselblatt M, Peters O, Thall PF, Kutluk T, Moghrabi A, Mahajan A, Rutkowski S, Diez B, Wang X, Pietsch T. Atypical choroid plexus papilloma: clinical experience in the CPT-SIOP-2000 study. J Neuro-Oncol. 2009;95:383–92.

25. Tabori U, Shlien A, Baskin B, Levitt S, Ray P, Alon N, Hawkins C, Bouffet E, Pienkowska M, Lafay-Cousin L, Gozali A. TP53 alterations determine clinical subgroups and survival of patients with choroid plexus tumors. J Clin Oncol. 2010;28(12):1995–2001.

26. Anderson M, Babington P, Taheri R, Diolombi M, Sherman JH. Unique presentation of cerebellopontine angle choroid plexus papillomas: case report and review of the literature. J Neurol Surg Rep. 2013;75:e27–32.

27. Verma SK, Satyarthee GD, Sharma BS. Giant choroid plexus papilloma of the lateral ventricle in fetus. J Pediatr Neurosci. 2014;9(2):185.

28. Prasad GL, Mahapatra AK. Case series of choroid plexus papilloma in children at uncommon locations and review of the literature. Surg Neurol Int. 2015;6:151.

29. Umredkar AA, Chhabra R, Bal A, Das A. Choroid plexus papilloma of the fourth ventricle in a septuagenarian. J Neurosci Rural Pract. 2012;3(03):402–4.

30. Safaee M, Oh MC, Bloch O, Sun MZ, Kaur G, Auguste KI, Tihan T, Parsa AT. Choroid plexus papillomas: advances in molecular biology and understanding of tumorigenesis. Neuro Oncol. 2013;15(3):255–67.

31. Thomas C, Metrock K, Kordes U, Hasselblatt M, Dhall G. Epigenetics impacts upon prognosis and clinical management of choroid plexus tumors. J Neurooncol. 2020;148:39–45.

32. Jaiswal AK, Jaiswal S, Sahu RN, Das KB, Jain VK, Behari S. Choroid plexus papilloma in children: diagnostic and surgical considerations. J Pediatr Neurosci. 2009;4(1):10.

33. Sethi D, Arora R, Garg K, Tanwar P. Choroid plexus papilloma. Asian J Neurosurg. 2017;12(01):139–41.

34. Mishra A, Srivastava C, Singh SK, Chandra A, Ojha BK. Choroid plexus carcinoma: case report and review of literature. J Pediatr Neurosci. 2012;7(1):71.

35. Sun MZ, Oh MC, Ivan ME, Kaur G, Safaee M, Kim JM, Phillips JJ, Auguste KI, Parsa AT. Current management of choroid plexus carcinomas. Neurosurg Rev. 2014;37:179–92.

36. Jeibmann A, Hasselblatt M, Gerss J, Wrede B, Egensperger R, Beschorner R, Hans VH, Rickert CH, Wolff JE, Paulus W. Prognostic implications of atypical histologic features in choroid plexus papilloma. J Neuropathol Exp Neurol. 2006;65(11):1069–73.

37. Louis DN, Perry A, Wesseling P, Brat DJ, Cree IA, Figarella-Branger D, Hawkins C, Ng HK, Pfister SM, Reifenberger G, Soffietti R. The 2021 WHO classification of tumors of the central nervous system: a summary. Neuro Oncol. 2021;23(8):1231–51.

38. Doğu H, Akdemir H, Çetin S. Lateral horizontal head position approach for the lateral and anterior third ventricles: a subependymoma clinical case and literature review. Asian J Neurosurg. 2022;17:642.

39. Lopes OG, Almeida FC, Cabral GA, Guimaraes RD, da Silva Filho RC, Landeiro JA. Intraparenchymal subependymoma: case report and literature review. Surg Neurol Int. 2021;12:154.

40. Mei G, Liu X, Zhou P, Shen M. Clinical and imaging features of subependymal giant cell astrocytoma: report of 20 cases. Chin Neurosurg J. 2017;3(02):67–73.

41. Rushing EJ, Cooper PB, Quezado M, Begnami M, Crespo A, Smirniotopoulos JG, Ecklund J, Olsen C, Santi M. Subependymoma revisited: clinicopathological evaluation of 83 cases. J Neurooncol. 2007;85:297–305.

42. Jain A, Amin AG, Jain P, Burger P, Jallo GI, Lim M, Bettegowda C. Subependymoma: clinical features and surgical outcomes. Neurol Res. 2012;34(7):677–84.

43. Wu J, Armstrong TS, Gilbert MR. Biology and management of ependymomas. Neuro Oncol. 2016;18(7):902–13.

44. Grajkowska W, Kotulska K, Jurkiewicz E, Roszkowski M, Daszkiewicz P, Jóźwiak S, Matyja E. Subependymal giant cell astrocytomas with atypical histological features mimicking malignant gliomas. Folia Neuropathol. 2011;49(1):39–46.

45. Almubarak AO, Abdullah J, Al Hindi H, AlShail E. Infantile atypical subependymal giant cell astrocytoma. Neurosci J. 2020;25(1):61–4.

46. Campen CJ, Porter BE. Subependymal giant cell astrocytoma (SEGA) treatment update. Curr Treat Options Neurol. 2011;13:380–5.

47. Stein JR, Reidman DA. Imaging manifestations of a subependymal giant cell astrocytoma in tuberous sclerosis. Case Rep Radiol. 2016;2016:3750450.

48. Orlov K, Gorbatykh A, Berestov V, Shayakhmetov T, Kislitsin D, Seleznev P, Strelnikov N. Superselective transvenous embolization with onyx and n-BCA for vein of galen aneurysmal malformations with restricted transarterial access: safety, efficacy, and technical aspects. Childs Nerv Syst. 2017;33:2003–10.

49. Park SJ, Jung TY, Kim SK, Lee KH. Tumor control of third ventricular central neurocytoma after gamma knife radiosurgery in an elderly patient: a case report and literature review. Medicine. 2018;97(50):e13657.

50. Lee SJ, Bui TT, Chen CH, Lagman C, Chung LK, Sidhu S, Seo DJ, Yong WH, Siegal TL, Kim M, Yang I. Central neurocytoma: a review of clinical management and histopathologic features. Brain Tumor Res Treat. 2016;4(2):49–57.

51. Chen CL, Shen CC, Wang J, Lu CH, Lee HT. Central neurocytoma: a clinical, radiological and pathological study of nine cases. Clin Neurol Neurosurg. 2008;110(2):129–36.

52. Cook DJ, Christie SD, Macaulay RJ, Rheaume DE, Holness RO. Fourth ventricular neurocytoma: case report and review of the literature. Can J Neurol Sci. 2004;31(4):558–64.

53. Choudhari KA, Kaliaperumal C, Jain A, Sarkar C, Soo MY, Rades D, Singh J. Central neurocytoma: a multi-disciplinary review. Br J Neurosurg. 2009;23(6):585–95.

54. Karnam S, Kottu R, Chowhan AK, Bodepati PC. Expression of p53 and, epidermal growth factor receptor in glioblastoma. Indian J Med Res. 2017;146(6):738.

55. Raguž M, Rotim A, Sajko T, Jurilj M, Splavski B, Rotim K. Mikrokirurško liječenje rijetkog slučajno nađenog intraventrikulskog meningeoma: prikaz slučaja i pregled relevantne literature. Acta Clin Croat. 2021;60(1):156–60.

56. Naeini RM, Yoo JH, Hunter JV. Spectrum of choroid plexus lesions in children. Am J Roentgenol. 2009;192(1):32–40.

57. Bhatoe HS, Singh P, Dutta V. Intraventricular meningiomas: a clinicopathological study and review of the literature. Neurosurg Focus. 2006;20(3):1–6.

58. Fernández CS, Lagarto EG, Rodríguez-Arias CA. Recurrent meningeal malignant tumor: assessment of differences in the solitary fibrous tumor/hemangiopericytoma spectrum through a case report. Neurocirugia. 2022;33(6):371–6.

59. Tüysüz B, Ercan-Sençicek AG, Özer E, Göç N, Yalçinkaya C, Bilguvar K. Severe phenotype in patients with X-linked hydrocephalus caused by a missense mutation in L1CAM. Turk Arch Pediatr. 2022;57(5):521.

60. Vaishnavathi V, Rangasami R, Suresh S, Suresh I. Fetal hemimegalencephaly. Neurol India. 2015;63(4):636.

61. Angulo-Maldonado M, Lara-Sarabia O, Cadena-Bonfanti A, Ulloa-Piza A. X-linked hereditary periventricular nodular heterotopia. Neurologia. 2021;37:232–4.

62. Koenig M, Dobyns WB, Di Donato N. Lissencephaly: update on diagnostics and clinical management. Eur J Pediatr Neurol. 2021;35:147–52.

63. Devisme L, Bouchet C, Gonzalès M, Alanio E, Bazin A, Bessières B, Bigi N, Blanchet P, Bonneau D, Bonnières M, Bucourt M. Cobblestone lissencephaly: neuropathological subtypes and correlations with genes of dystroglycanopathies. Brain. 2012;135(2):469–82.

64. Esenwa CC, Leaf DE. Colpocephaly in adults. BMJ Case Rep. 2013;2013:bcr2013009505.

65. Ansary A, Manjunatha CM, Ibhanesebhor S. Prenatal diagnosis of colpocephaly with absent corpus callosum. Scott Med J. 2015;60(1):e19–23.

66. Veerapaneni P, Veerapaneni KD, Yadala S. Schizencephaly. Treasure Island (FL): StatPearls Publishing; 2023.

67. Hoang VT, Hoang TH, Chansomphou V, Doan DT. Traumatic epidural hematoma in a patient with severe schizencephaly. Clin Case Rep. 2021;9(12):e05150.

68. Packard AM, Miller VS, Delgado MR. Schizencephaly: correlations of clinical and radiologic features. Neurology. 1997;48(5):1427–34.

69. Ramakrishnan S, Gupta V. Holoprosencephaly. Treasure Island (FL): StatPearls Publishing; 2023.

70. Bulakbasi N, Cancuri O, Kocaoğlu M. The middle interhemispheric variant of holoprosencephaly: magnetic resonance and diffusion tensor imaging findings. Br J Radiol. 2016;89(1063):20160115.

71. Das JM, Geetha R. Corpus callosum agenesis. Treasure Island (FL): StatPearls Publishing; 2023.

72. Brown WS, Paul LK. The neuropsychological syndrome of agenesis of the corpus callosum. J Int Neuropsychol Soc. 2019;25(3):324–30.

73. Page I, Gaillard F. Descriptive neuroradiology: beyond the hummingbird. Pract Neurol. 2020;20(6):463–71.

74. Shen O, Gelot AB, Moutard ML, Jouannic JM, Sela HY, Garel C. Abnormal shape of the cavum septi pellucidi: an indirect sign of partial agenesis of the corpus callosum. Ultrasound Obstetr Gynecol. 2015;46(5):595–9.

75. Webb EA, Dattani MT. Septo-optic dysplasia. Eur J Hum Genet. 2010;18(4):393–7.

76. Sandoval JI, De Jesus O. Hydranencephaly. Treasure Island (FL): StatPearls Publishing; 2023.

77. Pavone P, Praticò AD, Vitaliti G, Ruggieri M, Rizzo R, Parano E, Pavone L, Pero G, Falsaperla R. Hydranencephaly: cerebral spinal fluid instead of cerebral mantles. Ital J Pediatr. 2014;40:1–8.

78. Pokhraj PS, Jigar JP, Chetan M, Narottam AP. Congenital porencephaly in a new born child. J Clin Diagn Res. 2014;8(11):RJ01–2.

79. Oliván-Gonzalvo G, Calatayud-Maldonado V. Coexistence of central precocious puberty and intraventricular arachnoid cyst: a brief literature update. Iberoamerican J Med. 2021;3(1):65–70.

80. Ajtai B, Bertelson JA. Imaging of intracranial cysts. Continuum (Minneap Minn). 2016;22(5):1553–73.

81. Alluhaybi AA, Altuhaini K, Ahmad M. Fetal ventriculomegaly: a review of literature. Cureus. 2022;14(2):e22352.

82. Willems PJ, Brouwer OF, Dijkstra I, Wilmink J, Opitz JM, Reynolds JF. X-linked hydrocephalus. Am J Med Genet. 1987;27(4):921–8.

83. Altunkeser A, Kara T. Fetal ultrasound of type 2 and 3 Chiari malformation. Clin Res. 2018;4(2):1–3.

84. Kuhn J, Emmady PD. Chiari II malformation. Treasure Island (FL): StatPearls Publishing; 2023.

85. Kamitani T, Kawauchi Y, Kishida H, Kuroiwa Y. Ventriculardebris: MRI of ventriculitis. Neurology. 2003;60(9):1549.

86. Harris L, Munakomi S. Ventriculitis. Treasure Island (FL): StatPearls Publishing; 2023.

87. Jin SC, Ahn JS, Kwun BD, Kwon DH. Intraventricular cavernous malformation radiologically mimicking meningioma. J Korean Neurosurg Soc. 2008;44(5):345–7.

88. Mottolese C, Hermier M, Stan H, Jouvet A, Saint-Pierre G, Froment JC, Bret P, Lapras C. Central nervous system cavernomas in the pediatric age group. Neurosurg Rev. 2001;24(2–3):55–71.

89. Song WZ, Mao BY, Hu BF, Liu YH, Sun H, Mao Q. Intraventricular vascular malformations mimicking tumors: case reports and review of the literature. J Neurol Sci. 2008;266(1–2):63–9.

90. Batjer H, Samson D. Arteriovenous malformations of the posterior fossa: clinical presentation, diagnostic evaluation, and surgical treatment. J Neurosurg. 1986;64(6):849–56.

91. Romano N, Fiannacca M, Castaldi A. Imaging of ring-shaped lateral ventricular nodules (RSLVNs). Neurol Sci. 2023;44:2223–5.

92. Bertalanffy H, Mahmoodi R. Ventricular tumors. In: Winn HR, editor. Youmans and Winn neurological surgery. 7th ed. Philadelphia: Elsevier; 2017. p. 1192–221.

93. Oyama H, Noda S, Negoro M, Kinomoto T, Miyachi S, Kuwayama N, Kajita Y. Giant meningioma fed by the anterior choroidal artery: successful removal following embolization—case report. Neurol Med Chir. 1992;32(11):839–41.

94. Oertel JM, Schroeder HW, Gaab MR. Endoscopic stomy of the septum pellucidum: indications, technique, and results. Neurosurgery. 2009;64(3):482–93.

95. Nayar VV, Foroozan R, Weinberg JS, Yoshor D. Preservation of visual fields with the inferior temporal gyrus approach to the atrium: case report. J Neurosurg. 2009;110(4):740–3.

96. Patel TR, Gould GC, Baehring JM, Piepmeier JM. Surgical approaches to lateral and third ventricular tumors. In: Quinones-Hinojosa A, editor. Schmidek and sweet: operative neurosurgical techniques. 6th ed. Philadelphia: Saunders; 2012. p. 330–8.

97. Anderson RC, Ghatan S, Feldstein NA. Surgical approaches to tumors of the lateral ventricle. Neurosurg Clin N Am. 2003;14(4):509–25.

98. Yaşargil MG, Abdulrauf SI. Surgery of intraventricular tumors. Neurosurgery. 2008;62(6 suppl_3):1029–40.

99. D'Angelo VA, Galarza M, Catapano D, Monte V, Bisceglia M, Carosi I. Lateral ventricle tumors: surgical strategies according to tumor origin and development—a series of 72 cases. Neurosurgery. 2005;56(suppl_1):36–45.

100. Lawton MT, Golfinos JG, Spetzler RF. The contralateral transcallosal approach: experience with 32 patients. Neurosurgery. 1996;39(4):729–34.

101. Zagórska-Świeży K, Litwin JA, Gorczyca J, Pityński K, Miodoński AJ. Arterial supply and venous drainage of the choroid plexus of the human lateral ventricle in the prenatal period as revealed by vascular corrosion casts and SEM. Folia Morphol (Warsz). 2008;67(3):209–13.

102. Zhang WH, Xie M, Liu H, Wang X, Lin MH. Surgical challenges for lateral ventricle meningiomas: a consecutive series of 21 patients. J Huazhong Univ Sci Technol Med Sci. 2015;35:742–6.

103. Santoro A, Salvati M, Frati A, Polli FM. Surgical approaches to tumours of the lateral ventricles in the dominant hemisphere/comments. J Neurosurg Sci. 2002;46(2):60.

104. D'Angelo VA, Galarza M, Catapano D, Monte V, Bisceglia M, Carosi I. Lateral ventricle tumors: surgical strategies according to tumor origin and development—a series of 72 cases. Neurosurgery. 2008;62(6 Suppl 3):1066–75.

105. Barrenechea IJ, Márquez L, Miralles S, Baldoncini M, Peralta S. An alternative path to atrial lesions through a contralateral interhemispheric transfalcine transcingular infraprecuneus approach: a case report. Surg Neurol Int. 2020;11:407.

106. Asgari S, Engelhorn T, Brondics A, Sandalcioglu IE, Stolke D. Transcortical or transcallosal approach to ventricle-associated lesions: a clinical study on the prognostic role of surgical approach. Neurosurg Rev. 2003;26:192–7.

107. Ellenbogen RG. Transcortical surgery for lateral ventricular tumors. Neurosurg Focus. 2001;10(6):E2.

108. D'Ambrosio AL, Anderson RC, Bruce JN. Meningiomas of the lateral ventricle. Semin Neurosurg. 2003;14(3):237–44.

109. Danaila L. Primary tumors of the lateral ventricles of the brain. Chirurgia (Bucur). 2013;37:363–83.

110. Cikla U, Swanson KI, Tumturk A, Keser N, Uluc K, Cohen-Gadol A, Baskaya MK. Microsurgical resection of tumors of the lateral and third ventricles: operative corridors for difficult-to-reach lesions. J Neuro-Oncol. 2016;130:331–40.

111. Pendl G, Öztürk E, Haselsberger K. Surgery of tumours of the lateral ventricle. Acta Neurochir. 1992;116:128–36.

112. Tanaka Y, Sugita K, Kobayashi S, Takemae T, Hegde AS. Subdural fluid collections following transcortical approach to intra- or paraventricular tumours. Acta Neurochir. 1989;99(1–2):20–5.

113. Párraga RG. Surgical anatomy of the lateral ventricles check for updates Richard Gonzalo Párraga. In: Brain anatomy and neurosurgical approaches: a practical, illustrated, easy-to-use guide. Cham: Springer; 2023.

114. Yasargil MG. Microneurosurgery of CNS tumors. In: Yasargil MG, editor. Microneurosurgery, IVB. Stuttgart: Georg Thieme Verlag; 1996. p. 38–42.

115. Souweidane MM. Endoscopic surgery for intraventricular brain tumors in patients without hydrocephalus. Neurosurgery. 2005;57(4 Suppl):312–8.

116. Fernandes-Silva J, Silva SM, Alves H, Andrade JP, Arantes M. Neurosurgical anatomy of the floor of the third ventricle and related vascular structures. Surg Radiol Anat. 2021;43(12):1915–25.

117. Burbridge S, Stewart I, Placzek M. Development of the neuroendocrine hypothalamus. Compr Physiol. 2011;6(2):623–43.

118. Rhoton AL Jr. The lateral and third ventricles. Neurosurgery. 2002;51(4):S1–207.

119. Tsutsumi S, Ishii H, Ono H, Yasumoto Y. The third ventricle roof: an anatomical study using constructive interference in steady-state magnetic resonance imaging. Surg Radiol Anat. 2018;40:123–8.

120. Iacoangeli M, di Somma LG, Di Rienzo A, Alvaro L, Nasi D, Scerrati M. Combined endoscopic transforaminal–transchoroidal approach for the treatment of third ventricle colloid cysts. J Neurosurg. 2014;120(6):1471–6.

121. Hadzic A, Nguyen TD, Hosoyamada M, Tomioka NH, Bergersen LH, Storm-Mathisen J, Morland C. The lactate receptor HCA1 is present in the choroid plexus, the tela choroidea, and the neuroepithelial lining of the dorsal part of the third ventricle. Int J Mol Sci. 2020;21(18):6457.

122. Meyer FB, Bruce JN. Introduction to microsurgery of the third ventricle, pineal region, and tentorial incisura. Neurosurg Focus. 2016;40(video suppl 1):1.

123. Brzegowy K, Zarzecki MP, Musiał A, Aziz HM, Kasprzycki T, Tubbs RS, Popiela T, Walocha JA. The internal cerebral vein: new classification of branching patterns based on CTA. Am J Neuroradiol. 2019;40(10):1719–24.

124. Zohdi A, Elkheshin S. Endoscopic anatomy of the velum interpositum: a sequential descriptive anatomical study. Asian J Neurosurg. 2012;7(01):12–6.

125. Abdala-Vargas NJ, Cifuentes-Lobelo HA, Ordoñez-Rubiano E, Patiño-Gomez JG, Villalonga JF, Lucifero AG, Campero A, Forlizzi V, Baldoncini M, Luzzi S. Anatomic variations of the floor of the third ventricle: surgical implications for endoscopic third ventriculostomy. Surg Neurol Int. 2022;13:128.

126. Baka JJ, Spickler EM. Normal imaging anatomy of the suprasellar cistern and floor of the third ventricle. Semin Ultrasound CT MR. 1993;14(3):195–205.

127. Tsutsumi S, et al. The tuber cinereum as a circumventricular organ: an anatomical study using magnetic resonance imaging. Surg Radiol Anat. 2017;39(7):747–51.

128. Yamamoto I, Rhoton AL Jr, Peace DA. Microsurgery of the third ventricle: part 1: microsurgical anatomy. Neurosurgery. 1981;8(3):334–56.

129. Tubbs RS, Nguyen HS, Loukas M, Cohen-Gadol AA. Anatomic study of the lamina terminalis: neurosurgical relevance in approaching lesions within and around the third ventricle. Childs Nerv Syst. 2012;28:1149–56.

130. McKinley MJ, Congiu M, Denton DA, Park RG, Penschow J, Simpson JB, Tarjan E, Weisinger RS, Wright RD. The anterior wall of the third cerebral ventricle and homeostatic responses to dehydration. J Physiol. 1984;79(6):421–7.

131. Nugues MM, Roberts CM. Coral mortality and interaction with algae in relation to sedimentation. Coral Reefs. 2003;22:507–16.

132. Cossu M, Lubinu F, Orunesu G, Pau A, Viale ES, Sini MG, Turtas S. Subchoroidal approach to the third ventricle microsurgical anatomy. Surg Neurol. 1984;21(4):325–31.

133. Feletti A, Fiorindi A, Lavecchia V, Boscolo-Berto R, Marton E, Macchi V, De Caro R, Longatti P, Porzionato A, Pavesi G. A light on the dark side: in vivo endoscopic anatomy of the posterior third ventricle and its variations in hydrocephalus. J Neurosurg. 2020;135(1):309–17.

134. Simon E, Afif A, M'Baye M, Mertens P. Anatomy of the pineal region applied to its surgical approach. Neurochirurgie. 2015;61(2–3):70–6.

135. Quinones-Hinojosa A. Schmidek and sweet: operative neurosurgical techniques E-book: indications, methods and results (expert consult-online and print). Amsterdam: Elsevier Health Sciences; 2012.

136. Yamamoto I. Pineal region tumor: surgical anatomy and approach. J Neuro-Oncol. 2001;54:263–75.

137. Rhoton AL Jr. Tentorial incisura. Neurosurgery. 2000;47(suppl_3):S131–53.

138. Ono M, Ono M, Rhoton AL, Barry M. Microsurgical anatomy of the region of the tentorial incisura. J Neurosurg. 1984;60(2):365–99.

139. Figueiredo EG, Beer-Furlan A, Welling LC, Ribas EC, Schafranski M, Crawford N, Teixeira MJ, Rhoton AL Jr, Spetzler RF, Preul MC. Microsurgical approaches to the ambient cistern region: an anatomic and qualitative study. World Neurosurg. 2016;87:584–90.

140. Unsöld R, Ostertag CB, DeGroot J, Newton TH, Unsöld R, Ostertag CB, DeGroot J, Newton TH. Quadrigeminal cistern. Computer reformations of the brain and skull base. In: Anatomy and clinical application. Berlin: Springer; 1982. p. 152–62.

141. Matsuno H, Rhoton AL Jr, Peace D. Microsurgical anatomy of the posterior fossa cisterns. Neurosurgery. 1988;23(1):58–80.

142. Chaynes P. Microsurgical anatomy of the great cerebral vein of galen and its tributaries. J Neurosurg. 2003;99(6):1028–38.

143. Ahmed S, Javed G, Laghari A, Bareeqa S, Aziz K, Khan M, et al. Third ventricular tumors: a comprehensive literature review. Cureus. 2018;10:e3417.

144. Glastonbury CM, Osborn AG, Salzman KL. Masses and malformations of the third ventricle: normal anatomic relationships and differential diagnoses. Radiographics. 2011;31(7):1889–905.

145. Uygur E, Deniz B, Zafer K. Anterior third ventricle meningiomas. Report of two cases. Neurocirugía. 2008;19(4):356–60.

146. Castillo RG, Geise AW. Meningioma of the third ventricle. Surg Neurol. 1985;24(5):525–8.

147. Odake G, Tenjin H, Murakami N. Cyst of the choroid plexus in the lateral ventricle: case report and review of the literature. Neurosurgery. 1990;27(3):470–6.

148. Maeder PP, Holtås SL, Basibüyük LN, Salford LG, Tapper UA, Brun A. Colloid cysts of the third ventricle: correlation of MR and CT findings with histology and chemical analysis. Am J Roentgenol. 1990;155(1):135–41.

149. Ong APC, Frydenberg E, Rodriguez M, Chaganti J, Steel T. Unusual arachnoid cyst involving the third ventricle in an adult: case report, interdisciplinary. Neurosurgery. 2019;16:12–4.

150. Johannes W, Claudia G, Kasra S, Torben S, Volker H, Joachim G, et al. Chordoid glioma of the third ventricle: a systematic review and single-center experience. Interdiscip Neurosurg. 2019;18:100515.

151. Boyko OB, Curnes JT, Oakes WJ, Burger PC. Hamartomas of the tuber cinereum: CT, MR, and pathologic findings. AJNR Am J Neuroradiol. 1991;12:309–14.

152. Pawar S, Borde C, Patil A, Nagarkar R. Malignant epidermoid arising from the third ventricle: a case report. World J Radiol. 2019;11(5):74–80.

153. Suntharesan J, Gunasekara B, Atapattu N. Congenital hypopituitarism due to congenital craniopharyngioma: first case in Sri Lanka. Sri Lanka J Child Health. 2021;50(1):162–4.

154. Nowak A, Marchel A. Surgical treatment of intraventricular ependymomas and subependymomas. Neurol Neurochir Pol. 2012;46(4):333–43.

155. Lee SJ, Bui TT, Chen CHJ, Lagman C, Chung LK, Sidhu S, et al. Central neurocytoma: a review of clinical management and histopathologic features. Brain Tumor Res Treat. 2016;4(2):49.

156. Chan HS, Becker LE, Hoffman HJ, Humphreys RP, Hendrick BE, Fitz CR, Chuang SH. Myxopapillary ependymoma of the filum terminale and cauda equina in childhood: report of seven cases and review of the literature. Neurosurgery. 1984;14(2):204–10.

157. Soffietti R, Rudà R, Reardon D. Rare glial tumors. Handb Clin Neurol Gliomas. 2016;134:399–415.

158. Chen H, Zhou R, Liu J, Tang J. Central neurocytoma. J Clin Neurosci. 2012;19(6):849–53.

159. Smith AB, Smirniotopoulos JG, Horkanyne-Szakaly I. From the radiologic pathology archives: intraventricular neoplasms: radiologic–pathologic correlation. Radiographics. 2013;33(1):21–43. https://doi.org/10.1148/rg.331125192.

160. Yamamoto I, Rhoton AL, Peace DA. Microsurgery of the third ventricle: part 1. Neurosurgery. 1981;8:334–56.

161. Apuzzo MLJ, Chikovani OK, Gott PS, et al. Transcallosal, interforniceal approaches for lesions affecting the third ventricle: surgical considerations and consequences. Neurosurgery. 1982;10:547–54.

162. Kondziolka D, Lunsford LD. Stereotactic management of colloid cysts: factors predicting success. J Neurosurg. 1991;75:45–51.

163. Kochanski RB, Dawe R, Kocak M, Sani S. Identification of stria medullaris fibers in the massa intermedia using diffusion tensor imaging. World Neurosurg. 2018;112:e497–504.

164. El Damaty A, Langner S, Schroeder HWS. Ruptured massa intermedia secondary to hydrocephalus. World Neurosurg. 2017;97:749.e7–749.e10.

165. Pradilla G, Jallo G. Arachnoid cysts: case series and review of the literature. Neurosurg Focus. 2007;22(2):1–4.

166. Decq P, Brugières P, Le Guerinel C, Djindjian M, Kéravel Y, Nguyen JP. Percutaneous endoscopic treatment of suprasellar arachnoid cysts: ventriculocystostomy or ventriculocystocisternostomy? J Neurosurg. 1996;84(4):696–701.

167. Kalani MYS, Zabramski JM, Martirosyan NL, Spetzler RF. Developmental venous anomaly, capillary telangiectasia, cavernous malformation, and arteriovenous malformation: spectrum of a common pathological entity? Acta Neurochir. 2016;158(3):547–50.

168. Woodall MN, Catapano JS, Lawton MT, Spetzler RF. Cavernous malformations in and around the third ventricle: indications, approaches, and outcomes. Oper Neurosurg. 2020;18(6):736–46.

169. Cinalli G, Spennato P, Nastro A, Aliberti F, Trischitta V, Ruggiero C, Mirone G, Cianciulli E. Hydrocephalus in aqueductal stenosis. Childs Nerv Syst. 2011;27:1621–42.

170. Hirsch JF, Hirsch E, Sainte Rose C, Renier D, Pierre-Khan A. Stenosis of the aqueduct of sylvius. Etiology and treatment. J Neurosurg Sci. 1986;30(1–2):29–39.

171. Jellinger G. Anatomopathology of non-tumoral aqueductal stenosis. J Neurosurg Sci. 1986;30(1–2):1–16.

172. Kulkarni AV, Drake JM, Kestle JRW, Mallucci CL, Sgouros S, Constantini S. Endoscopic third ventriculostomy vs cerebrospinal fluid shunt in the treatment of hydrocephalus in children. Neurosurgery. 2010;67(3):588–93.

173. Du Boulay G, O'connell J, Currie J, Bostick T, Verity P. Further investigations on pulsatile movements in the cerebrospinal fluid pathways. Acta Radiol Diagn. 1972;13(P2):496–523.

174. Cinalli G, Spennato P, Cianciulli E, D'Armiento M. Hydrocephalus and aqueductal stenosis. In: Cinalli G, Maixner W, Saint-Rose C, editors. Pediatric hydrocephalus. Milan: Springer; 2004. p. 79–293.

175. Miše B, Klarica M, Seiwerth S, Bulat M. Experimental hydrocephalus and hydromyelia: a new insight in mechanism of their development. Acta Neurochir. 1996;138(7):862–9.

176. Greitz D. Radiological assessment of hydrocephalus: new theories and implications for therapy. Neurosurg Rev. 2004;27(3):145–65.

177. Spennato P, Tazi S, Bekaert O, Cinalli G, Decq P. Endoscopic third ventriculostomy for idiopathic aqueductal stenosis. World Neurosurg. 2013;79(2):S21–e13.

178. Jones RF, Stening WA, Brydon M. Endoscopic third ventriculostomy. Neurosurgery. 1990;26:86–91.

179. Emery SP, Narayanan S, Greene S. Fetal aqueductal stenosis: prenatal diagnosis and intervention. Prenat Diagn. 2020;40(1):58–65.

180. Emery SP, Hogge WA, Hill LM. Accuracy of prenatal diagnosis of isolated aqueductal stenosis. Prenat Diagn. 2015;35(4):319–24.

181. Rault E, Lacalm A, Massoud M, Massardier J, Di Rocco F, Gaucherand P, et al. The many faces of prenatal imaging diagnosis of primitive aqueduct obstruction. Eur J Paediatr Neurol. 2018;22(6):910–8.

182. Kulkarni AV, Sgouros S, Constantini S, Investigators IIHS. International infant hydrocephalus study: initial results of a prospective, multicenter comparison of endoscopic third ventriculostomy (ETV) and shunt for infant hydrocephalus. Childs Nerv Syst. 2016;32(6):1039–48.

183. Buxton N. Changes in third ventricular size with neuroendoscopic third ventriculostomy: a blinded study. J Neurol Neurosurg Psychiatry. 2002;72(3):385–7.

184. Kulkarni AV, Drake JM, Kestle JRW, Mallucci CL, Sgouros S, Constantini S. Predicting who will benefit from endoscopic third ventriculostomy compared with shunt insertion

in childhood hydrocephalus using the ETV success score. J Neurosurg Pediatr. 2010;6(4):310–5.

185. Feng Z, Li Q, Gu J, Shen W. Update on endoscopic third ventriculostomy in children. Pediatr Neurosurg. 2018;53(6):367–70.

186. Munda M, Spazzapan P, Bosnjak R, Velnar T. Endoscopic third ventriculostomy in obstructive hydrocephalus: a case report and analysis of operative technique. World J Clin Cases. 2020;8(14):3039–49.

187. Kulkarni AV, Riva-cambrin J, Holubkov R, Browd SR, Cochrane DD, Drake JM, et al. Endoscopic third ventriculostomy in children: prospective, multicenter results from the hydrocephalus clinical research network. J Neurosurg Pediatr. 2016;18:423–9.

188. Souweidane MM, Morgenstern PF, Kang S, Tsiouris AJ, Roth J. Endoscopic third ventriculostomy in patients with a diminished prepontine interval. J Neurosurg Pediatr. 2010;5(3):250–4.

189. Kombogiorgas D, Sgouros S. Assessment of the influence of operative factors in the success of endoscopic third ventriculostomy in children. Childs Nerv Syst. 2006;22(10):1256–62.

190. Froelich SC, Abdel Aziz KM, Cohen PD, van Loveren HR, Keller JT. Microsurgical and endoscopic anatomy of Liliequist's membrane: a complex and variable structure of the basal cisterns. Neurosurgery. 2008;63(1 Suppl 1):ONS1–8.

191. Weissenberger AA, Dell ML, Liow K, Theodore W, Frattali CM, Hernandez D, et al. Aggression and psychiatric comorbidity in children with hypothalamic hamartomas and their unaffected siblings. J Am Acad Child Adolesc Psychiatry. 2001;40(6):696–703.

192. Chapman KE, Kim DY, Rho JM, Ng YT, Kerrigan JF. Ketogenic diet in the treatment of seizures associated with hypothalamic hamartomas. Epilepsy Res. 2011;94(3):218–21.

193. Blumberg J, Fernández IS, Vendrame M, Oehl B, Tatum WO, Schuele S, et al. Dacrystic seizures: demographic, semiologic, and etiologic insights from a multicenter study in long-term video-EEG monitoring units. Epilepsia. 2012;53(10):1810–9.

194. Loiseau P, Cohadon P, Cohadon S. Gelastic epilepsy a review and report of five cases. Epilepsia. 1971;12(4):313–23.

195. Alomari SO, Houshiemy MNE, Bsat S, Moussalem CK, Allouh M, Omeis IA. Hypothalamic hamartomas: a comprehensive review of the literature—part 1: neurobiological features, clinical presentations and advancements in diagnostic tools. Clin Neurol Neurosurg. 2020;197:106076.

196. Harrison VS, Oatman O, Kerrigan JF. Hypothalamic hamartoma with epilepsy: review of endocrine comorbidity. Epilepsia. 2017;58:50–9.

197. Ghanta R, Kongara S, Koti K, Meher G. Surgical excision of hypothalamic hamartoma in a 20 months old boy with precocious puberty. Indian J Endocrinol Metab. 2011;15(7):255.

198. Mittal S, Mittal M, Montes JL, Farmer JP, Andermann F. Hypothalamic hamartomas. Part 2. Surgical considerations and outcome. Neurosurg Focus. 2013;34(6):E7.

199. Alomari SO, El Houshiemy MN, Bsat S, Moussalem CK, Allouh M, Omeis IA. Hypothalamic hamartomas: a comprehensive review of literature—part 2: medical and surgical management update. Clin Neurol Neurosurg. 2020;195:106074.

200. Ng Y, Rekate HL, Prenger EC, Chung SS, Feiz-Erfan I, Wang NC, et al. Transcallosal resection of hypothalamic hamartoma for intractable epilepsy. Epilepsia. 2006;47(7):1192–202.

201. Régis J, Scavarda D, Tamura M, Nagayi M, Villeneuve N, Bartolomei F, et al. Epilepsy related to hypothalamic hamartomas: surgical management with special reference to gamma knife surgery. Childs Nerv Syst. 2006;22(8):881–95.

202. Cavallo LM, Di Somma A, de Notaris M, Prats-Galino A, Aydin S, Catapano G, Solari D, de Divitiis O, Somma T, Cappabianca P. Extended endoscopic endonasal approach to the third ventricle: multimodal anatomical study with surgical implications. World Neurosurg. 2015;84(2):267–78.

203. Adeolu AA, Osazuwa UA, Oremakinde AA, Oyemolade TA, Shokunbi MT. Combined microsurgical extra-axial and transcortical transventricular endoscopic excision of parasellar tumors with ventricular extension. Ann Afr Med. 2015;14(3):155.

204. Shirane R, Kumabe T, Yoshida Y, Su CC, Jokura H, Umezawa K, Yoshimoto T. Surgical treatment of posterior fossa tumors via the occipital transtentorial approach: evaluation of operative safety and results in 14 patients with anterosuperior cerebellar tumors. J Neurosurg. 2001;94(6):927–35.

205. Patel P, Cohen-Gadol AA, Boop F, Klimo P. Technical strategies for the transcallosal transforaminal approach to third ventricle tumors: expanding the operative corridor. J Neurosurg Pediatr. 2014;14(4):365–71.

206. Knaus H, Matthias S, Koch A, Thomale UW. Single burr hole endoscopic biopsy with third ventriculostomy—measurements and computer-assisted planning. Childs Nerv Syst. 2011;27(8):1233–41.

207. Gielen FL, Hertel F, Gemmar P, Appenrodt P, Inventors, Medtronic Inc, Assignee. Method for planning a surgical procedure. United States patent US 8160676. 2012.

208. Behari S, Jaiswal S, Nair P, Garg P, Jaiswal AK. Tumors of the posterior third ventricular region in pediatric patients: the Indian perspective and a review of literature. J Pediatr Neurosci. 2011;6(Suppl 1):S56.

209. Zoli M, Sambati L, Milanese L, Foschi M, Faustini-Fustini M, Marucci G, de Biase D, Tallini G, Cecere A, Mignani F, Sturiale C. Postoperative outcome of body core temperature rhythm and sleep-wake cycle in third ventricle craniopharyngiomas. Neurosurg Focus. 2016;41(6):E12.

210. Danaila L, Radoi M. Surgery of tumors of the third ventricle region. Chirurgia. 2013;108(4):456–62.

211. Symss NP, Ramamurthi R, Rao SM, Vasudevan MC, Jain PK, Pande A. Management outcome of the transcallosal, transforaminal approach to colloid cysts of the anterior third ventricle: an analysis of 78 cases. Neurol India. 2011;59(4):542.

212. Chamoun R, Couldwell WT. Transcortical–transforaminal microscopic approach for purely intraventricular craniopharyngioma. Neurosurg Focus. 2013;34(1 Suppl):1.

213. Nair S, Gopalakrishnan CV, Menon G, Easwer HV, Abraham M. Interhemispheric transcallosal transforaminal approach and its variants to colloid cyst of third ventricle: technical issues based on a single institutional experience of 297 cases. Asian J Neurosurg. 2016;11(3):292.

214. Kim MH. Transcortical endoscopic surgery for intraventricular lesions. J Korean Neurosurg Soc. 2017;60(3):327.

215. Peltier J, Roussel M, Gerard Y, Lassonde M, Deramond H, Le Gars D, De Beaumont L, Godefroy O. Functional consequences of a section of the anterior part of the body of the corpus callosum: evidence from an interhemispheric transcallosal approach. J Neurol. 2012;259(9):1860–7.

216. Hicdonmez T, Hamamcioglu MK, Parsak T, Cukur Z, Cobanoglu S. A laboratory training model for interhemispheric-transcallosal approach to the lateral ventricle. Neurosurg Rev. 2006;29(2):159–62.

217. Mazza M, Di Rienzo A, Costagliola C, Roncone R, Casacchia M, Ricci A, Galzio RJ. The interhemispheric transcallosal–transversal approach to the lesions of the anterior and middle third ventricle: surgical validity and neuropsychological evaluation of the outcome. Brain Cogn. 2004;55(3):525–34.

218. Komura S, Akiyama Y, Suzuki H, Yokoyama R, Mikami T, Mikuni N. Far-anterior interhemispheric transcallosal approach for a central neurocytoma in the lateral ventricle. Neurol Med Chir. 2019;59:511.

219. Belykh E, Yağmurlu K, Lei T, Safavi-Abbasi S, Oppenlander ME, Martirosyan NL, Byvaltsev VA, Spetzler RF, Nakaji P, Preul MC. Quantitative anatomical comparison of the ipsilateral and contralateral interhemispheric transcallosal approaches to the lateral ventricle. J Neurosurg. 2018;128(5):1492–502.

220. Aryan HE, Ozgur BM, Jandial R, Levy ML. Complications of interhemispheric transcallosal approach in children: review of 15 years experience. Clin Neurol Neurosurg. 2006;108(8):790–3.

221. Winkler PA, Ilmberger J, Krishnan KG, Reulen HJ. Transcallosal interforniceal-transforaminal approach for removing lesions occupying the third ventricular space: clinical and neuropsychological results. Neurosurgery. 2000;46(4):879–90.

222. Erturk ME, Kayalioglu G, Ozer MA, Ozgur T. Morphometry of the anterior third ventricle region as a guide for the transcallosal-interforniceal approach. Neurol Med Chir. 2004;44(6):288–93.

223. Bozkurt B, Yağmurlu K, Belykh E, Meybodi AT, Staren MS, Aklinski JL, Preul MC, Grande AW, Nakaji P, Lawton MT. Quantitative anatomic analysis of the transcallosal–transchoroidal approach and the transcallosal-subchoroidal approach to the floor of the third ventricle: an anatomic study. World Neurosurg. 2018;118:219–29.

224. Hosainey SA, Meling TR. A 34-year-old woman with brainstem cavernous malformation: the anterior transcallosal transchoroidal approach and literature review. J Neurol Surg Rep. 2014;75(2):e236.

225. Ulm AJ, Russo A, Albanese E, Tanriover N, Martins C, Mericle RM, Pincus D, Rhoton AL. Limitations of the transcallosal transchoroidal approach to the third ventricle. J Neurosurg. 2009;111(3):600–9.

226. Hassaneen W, Suki D, Salaskar AL, Levine NB, DeMonte F, Lang FF, McCutcheon IE, Dorai Z, Feiz-Erfan I, Wildrick DM, Sawaya R. Immediate morbidity and mortality associated with transcallosal resection of tumors of the third ventricle. J Clin Neurosci. 2010;17(7):830–6.

227. Krishna V, Blaker B, Kosnik L, Patel S, Vandergrift W. Trans-lamina terminalis approach to third ventricle using supraorbital craniotomy: technique description and literature review for outcome comparison with anterior, lateral and trans-sphenoidal corridors. Minim Invasive Neurosurg. 2011;54(05/06):236–42.

228. Amagasa M, Ishibashi Y, Kayama T, Suzuki J. A total removal case of cavernous angioma at the lateral wall of the third ventricle with interhemispheric trans-lamina terminalis approach. No shinkei geka Neurol Surg. 1984;12(4):517–22.

229. Hori T, Kawamata T, Amano K, Aihara Y, Ono M, Miki N. Anterior interhemispheric approach for 100 tumors in and around the anterior third ventricle. Oper Neurosurg. 2010;66(suppl_1):65–75.

230. Aftahy AK, Barz M, Wagner A, Liesche-Starnecker F, Negwer C, Meyer B, Gempt J. The interhemispheric fissure—surgical outcome of interhemispheric approaches. Neurosurg Rev. 2020;27:1–2.

231. Liu JK, Christiano LD, Gupta G, Carmel PW. Surgical nuances for removal of retrochiasmatic craniopharyngiomas via the transbasal subfrontal translamina terminalis approach. Neurosurg Focus. 2010;28(4):E6.

232. Dehdashti AR, de Tribolet N. Frontobasal interhemispheric trans-lamina terminalis approach for suprasellar lesions. Oper Neurosurg. 2005;56(suppl_4):418.

233. Weil AG, Robert T, Alsaiari S, Obaid S, Bojanowski MW. Using the trans-lamina terminalis route via a pterional approach to resect a retrochiasmatic craniopharyngioma involving the third ventricle. Neurosurg Focus. 2016;40(video suppl 1):1.

234. Hendricks BK, Cohen-Gadol AA. Resection of large recurrent third ventricular epidermoid tumors through the posterior interhemispheric transcallosal approach. Neurosurg Focus. 2016;40(video suppl 1):1.

235. Patel PG, Cohen-Gadol AA, Mercier P, Boop FA, Klimo P Jr. The posterior transcallosal approach to the pineal region and posterior third ventricle: intervenous and paravenous variants. Oper Neurosurg. 2017;13(1):77–88.

236. Hersh DS, Sanford KN, Boop FA. Posterior transcallosal intervenous-interforniceal approach to a periaqueductal tumor. Neurosurg Focus Video. 2019;1(2):V8.

237. Moshel YA, Parker EC, Kelly PJ. Occipital transtentorial approach to the precentral cerebellar fissure and posterior incisural space. Neurosurgery. 2009;65(3):554–64.

238. Kawashima M, Rhoton AL Jr, Matsushima T. Comparison of posterior approaches to the posterior incisural space: microsurgical anatomy and proposal of a new method, the occipital bi-transtentorial/falcine approach. Neurosurgery. 2002;51(5):1208–21.

239. Iwami K, Fujii M, Saito K. Occipital transtentorial/falcine approach, a "cross-court" trajectory to accessing contralateral posterior thalamic lesions: case report. J Neurosurg. 2016;127(1):165–70.

240. Ma Y, Lan Q. An anatomic study of the occipital transtentorial keyhole approach. World Neurosurg. 2013;80(1–2):183–9.

241. Yoshimoto K, Araki Y, Amano T, Matsumoto K, Nakamizo A, Sasaki T. Clinical features and pathophysiological mechanism of the hemianoptic complication after the occipital transtentorial approach. Clin Neurol Neurosurg. 2013;115(8):1250–6.

242. Konovalov AN, Pitskhelauri DI. Infratentorial supracerebellar approach to the colloid cysts of the third ventricle. Neurosurgery. 2001;49(5):1116–23.

243. Cardia A, Caroli M, Pluderi M, Arienta C, Gaini SM, Lanzino G, Tschabitscher M. Endoscope-assisted infratentorial–supracerebellar approach to the third ventricle: an anatomical study. J Neurosurg Pediatr. 2006;104(6):409–14.

244. Aboul-Enein H, Sabry AA, Farhoud AH. Supracerebellar infratentorial approach with paramedian expansion for posterior third ventricular and pineal region lesions. Clin Neurol Neurosurg. 2015;139:100–9.

245. Ammirati M, Bernardo A, Musumeci A, Bricolo A. Comparison of different infratentorial—supracerebellar approaches to the posterior and middle incisural space: a cadaveric study. J Neurosurg. 2002;97(4):922–8.

246. Laborde G, Gilsbach JM, Harders A, Seeger W. Experience with the infratentorial supracerebellar approach in lesions of the quadrigeminal region, posterior third ventricle, culmen cerebelli, and cerebellar peduncle. Acta Neurochir. 1992;114(3–4):135–8.

247. Jakola AS, Bartek J, Mathiesen T. Venous complications in supracerebellar infratentorial approach. Acta Neurochir. 2013;155(3):477–8.

248. Bergsneider M, Holly LT, Lee JH, King WA, Frazee JG. Endoscopic management of cysticercal cysts within the lateral and third ventricles. J Neurosurg. 2000;92(1):14–23.

249. Hussein RA, Al-Jumaily MH, Abdulhameed AA, Sulaiman II. Collod cyst; endoscopic versus transcallusal approach. NeuroQuantology. 2020;18(1):8.

250. Goel RK, Ahmad FU, Vellimana AK, Suri A, Chandra PS, Kumar R, Sharma BS, Mahapatra AK. Endoscopic management of intraventricular neurocysticercosis. J Clin Neurosci. 2008;15(10):1096–101.

251. Hellwig D, Grotenhuis JA, Tirakotai W, Riegel T, Schulte DM, Bauer BL, Bertalanffy H. Endoscopic third ventriculostomy for obstructive hydrocephalus. Neurosurg Rev. 2005;28(1):1–34.

252. Horn EM, Feiz-Erfan I, Bristol RE, Lekovic GP, Goslar PW, Smith KA, Nakaji P, Spetzler RF. Treatment options for third ventricular colloid cysts: comparison of open microsurgical versus endoscopic resection. Neurosurgery. 2007;60(4):613–20.

253. Duffner F, Schiffbauer H, Glemser D, Skalej M, Freudenstein D. Anatomy of the cerebral ventricular system for endoscopic neurosurgery: a magnetic resonance study. Acta Neurochir. 2003;145(5):359–68.

254. Chibbaro S, Di Rocco F, Makiese O, Reiss A, Poczos P, Mirone G, Servadei F, George B, Crafa P, Polivka M, Romano A. Neuroendoscopic management of posterior third ventricle and pineal region tumors: technique, limitation, and possible complication avoidance. Neurosurg Rev. 2012;35(3):331–40.

255. Desai KI, Nadkarni TD, Muzumdar DP, Goel AH. Surgical management of colloid cyst of the third ventricle—a study of 105 cases. Surg Neurol. 2002;57(5):295–302.

256. Kobayashi T, Tsugawa T, Hashizume C, Arita N, Hatano H, Iwami K, Nakazato Y, Mori Y. Therapeutic approach to chordoid glioma of the third ventricle. Neurol Med Chir. 2013;53(4):249–55.

257. Green BT, Tendler DA. Cerebral air embolism during upper endoscopy: case report and review. Gastrointest Endosc. 2005;61(4):620–3.

258. Mirski MA, Lele AV, Fitzsimmons L, Toung TJ, Warltier DC. Diagnosis and treatment of vascular air embolism. J Am Soc Anesthesiol. 2007;106(1):164–77.

259. Blanc P, Boussuges A, Henriette K, Sainty J, Deleflie M. Iatrogenic cerebral air embolism: importance of an early hyperbaric oxygenation. Intensive Care Med. 2002;28(5):559–63.

260. Elwatidy SM, Albakr AA, Al Towim AA, Malik SH. Tumors of the lateral and third ventricle: surgical management and outcome analysis in 42 cases. Neurosciences. 2017;22(4):274.

261. Milligan BD, Meyer FB. Morbidity of transcallosal and transcortical approaches to lesions in and around the lateral and third ventricles: a single-institution experience. Neurosurgery. 2010;67(6):1483–96.

262. Villani R, Papagno C, Tomei G, Grimoldi N, Spagnoli D, Bello L. Transcallosal approach to tumors of the third ventricle. Surgical results and neuropsychological evaluation. J Neurosurg Sci. 1997;41(1):41.

263. Chernov MF, Kamikawa S, Yamane F, Ishihara S, Kubo O, Hori T. Neurofiberscopic biopsy of tumors of the pineal region and posterior third ventricle: indications, technique, complications, and results. Neurosurgery. 2006;59(2):267–77.

264. Mercier P, Bernard F, Delion M. Microsurgical anatomy of the fourth ventricle. Neurochirurgie. 2021;67(1):14–22.

265. Matsushima T, Rhoton AL, Lenkey C. Microsurgery of the fourth ventricle: part 1: microsurgical anatomy. Neurosurgery. 1982;11(5):631–67.

266. Salma A, Yeremeyeva E, Baidya NB, Sayers MP, Ammirati M. An endoscopic, cadaveric analysis of the roof of the fourth ventricle. J Clin Neurosci. 2013;20(5):710–4.

267. Baran O, Baydin S, Mirkhasilova M, Bayramli N, Bilgin B, Middlebrooks E, Ozlen F, Tanriover N. Microsurgical anatomy and surgical exposure of the cerebellar peduncles. Neurosurg Rev. 2022 Jun;45:2095–117.

268. Matsushima K, Yagmurlu K, Kohno M, Rhoton AL. Anatomy and approaches along the cerebellar-brainstem fissures. J Neurosurg. 2016;124(1):248–63.

269. Tubbs RS, Bosmia AN, Loukas M, Hattab EM, Cohen-Gadol AA. The inferior medullary velum: anatomical study and neurosurgical relevance. J Neurosurg. 2013;118(2):315–8.

270. Mussi AC, Rhoton AL. Telovelar approach to the fourth ventricle: microsurgical anatomy. J Neurosurg. 2000;92(5):812–23.

271. Matsushima T, Rutka J, Matsushima K. Evolution of cerebellomedullary fissure opening: its effects on posterior fossa surgeries from the fourth ventricle to the brainstem. Neurosurg Rev. 2021;44:699–708.

272. Bentson JR, Alberti JB. Lateral recesses of the fourth ventricle. Radiology. 1972;104(3):593–9.

273. Bogucki J, Gielecki J, Czernicki Z. The anatomical aspects of a surgical approach through the floor of the fourth ventricle. Acta Neurochir. 1997;139:1014–9.

274. Rutledge C, Tonetti DA, Raygor KP, Abla AA. How I do it: horizontal fissure approach to the middle cerebellar peduncle. Acta Neurochir. 2022;164:763–6.

275. Mennink LM, Van Dijk JM, Van Der Laan BF, Metzemaekers JD, Van Laar PJ, Van Dijk P. The relation between flocculus volume and tinnitus after cerebellopontine angle tumor surgery. Hear Res. 2018;361:113–20.

276. Tubbs RS, Shoja MM, Aggarwal A, Gupta T, Loukas M, Sahni D, Ansari SF, Cohen-Gadol AA. Choroid plexus of the fourth ventricle: review and anatomic study highlighting anatomical variations. J Clin Neurosci. 2016;26:79–83.

277. Oberman DZ, Baldoncini M, Campero A, Ajler P. Surgical anatomy of the fourth ventricle surgical anatomy of the fourth ventricle. In: Brain anatomy and neurosurgical approaches: a practical, illustrated, easy-to-use guide. Cham: Springer International Publishing; 2023. p. 391–402.

278. Benner D, Hendricks BK, Benet A, Graffeo CS, Scherschinski L, Srinivasan VM, Catapano JS, Lawrence PM, Schornak M, Lawton MT. A system of anatomical triangles defining dissection routes to brainstem cavernous malformations: definitions and application to a cohort of 183 patients. J Neurosurg. 2022;138:768.

279. Lucifero AG, Baldoncini M, Bruno N, Tartaglia N, Ambrosi A, Marseglia GL, Galzio R, Campero A, Hernesniemi J, Luzzi S. Microsurgical neurovascular anatomy of the brain: the posterior circulation (part II). Acta Biomed. 2021;92(Suppl 4):e2021413.

280. Adigun OO, Al-Dhahir MA. Anatomy, head and neck, cerebrospinal fluid. Treasure Island (FL): StatPearls Publishing; 2019.

281. Hoffman S, Propp JM, McCarthy BJ. Temporal trends in incidence of primary brain tumors in the United States, 1985–1999. Neuro-Oncology. 2006;8(1):27–37.

282. Packer RJ, Sutton LN, Rorke LB, Littman PA, Sposto R, Rosenstock JG, Bruce DA, Schut L. Prognostic importance of cellular differentiation in medulloblastoma of childhood. J Neurosurg. 1984;61(2):296–301.

283. Louis DN, Ohgaki H, Wiestler OD, Cavenee WK, Burger PC, Jouvet A, Scheithauer BW, Kleihues P. The 2007 WHO classification of tumours of the central nervous system. Acta Neuropathol. 2007;114(2):97–109.

284. McManamy CS, Pears J, Weston CL, Hanzely Z, Ironside JW, Taylor RE, Grundy RG, Clifford SC, Ellison DW, Clinical Brain Tumour Group, Children's Cancer and Leukaemia Group (formerly the UK Children's Cancer Study Group), UK. Nodule formation and desmoplasia in medulloblastomas—defining the nodular/desmoplastic variant and its biological behavior. Brain Pathol. 2007;17(2):151–64.

285. Hooper CM, Hawes SM, Kees UR, Gottardo NG, Dallas PB. Gene expression analyses of the spatio-temporal relationships of human medulloblastoma subgroups during early human neurogenesis. PLoS One. 2014;9(11):e112909.

286. Northcott PA, Korshunov A, Witt H, Hielscher T, Eberhart CG, Mack S, Bouffet E, Clifford SC, Hawkins CE, French P, Rutka JT. Medulloblastoma comprises four distinct molecular variants. J Clin Oncol. 2011;29(11):1408.

287. Kool M, Korshunov A, Remke M, Jones DT, Schlanstein M, Northcott PA, Cho YJ, Koster J, Schouten-van Meeteren A, Van Vuurden D, Clifford SC. Molecular subgroups of medulloblastoma: an international meta-analysis of transcriptome, genetic aberrations, and clinical data of WNT, SHH, group 3, and group 4 medulloblastomas. Acta Neuropathol. 2012;123(4):473–84.

288. Cho YJ, Tsherniak A, Tamayo P, Santagata S, Ligon A, Greulich H, Berhoukim R, Amani V, Goumnerova L, Eberhart CG, Lau CC. Integrative genomic analysis of medulloblastoma identifies a molecular subgroup that drives poor clinical outcome. J Clin Oncol. 2011;29(11):1424.

289. Clevers H. Wnt/β-catenin signaling in development and disease. Cell. 2006;127(3):469–80.

290. Padovani L, Sunyach MP, Perol D, Mercier C, Alapetite C, Haie-Meder C, Hoffstetter S, Muracciole X, Kerr C, Wagner JP, Lagrange JL. Common strategy for adult and pediatric medulloblastoma: a multicenter series of 253 adults. Int J Radiat Oncol Biol Phys. 2007;68(2):433–40.

291. Bourgouin PM, Tampieri D, Grahovac SZ, Léger C, Del Carpio R, Melançon D. CT and MR imaging findings in adults with cerebellar medulloblastoma: comparison with findings in children. AJR Am J Roentgenol. 1992;159(3):609–12.

292. Loree J, Mehta V, Bhargava R. Cranial magnetic resonance imaging findings of leptomeningeal contrast enhancement after pediatric posterior fossa tumor resection and its significance: case report. J Neurosurg Pediatr. 2010;6(1):87–91.

293. Massimino M, Antonelli M, Gandola L, Miceli R, Pollo B, Biassoni V, Schiavello E, Buttarelli FR, Spreafico F, Collini P, Giangaspero F. Histological variants of medulloblastoma are the most powerful clinical prognostic indicators. Pediatr Blood Cancer. 2013;60(2):210–6.

294. Chang CH, Housepian EM, Herbert C. An operative staging system and a megavoltage radiotherapeutic technic for cerebellar medulloblastomas. Radiology. 1969;93(6):1351–9.

295. Müller K, Mynarek M, Zwiener I, Siegler N, Zimmermann M, Christiansen H, Budach W, Henke G, Warmuth-Metz M, Pietsch T, von Hoff K. Postponed is not canceled: role of craniospinal radiation therapy in the management of recurrent infant medulloblastoma—an experience from the HIT-REZ 1997 and, 2005 studies. Int J Radiat Oncol Biol Phys. 2014;88(5):1019–24.

296. Armstrong TS, Vera-Bolanos E, Bekele BN, Aldape K, Gilbert MR. Adult ependymal tumors: prognosis and the M. D. Anderson Cancer Center experience. Neuro-Oncology. 2010;12(8):862–70.

297. Fuller CE. Ependymomas and choroid plexus tumors. In: Practical surgical neuropathology: a diagnostic approach. Amsterdam: Elsevier; 2018. p. 145–69.

298. Rosenblum MK. Ependymal tumors: a review of their diagnostic surgical pathology. Pediatr Neurosurg. 1998;28(3):160–5.

299. Witt H, Mack SC, Ryzhova M, Bender S, Sill M, Isserlin R, Benner A, Hielscher T, Milde T, Remke M, Jones DT. Delineation of two clinically and molecularly distinct subgroups of posterior fossa ependymoma. Cancer Cell. 2011;20(2):143–57.

300. Lamszus K, Lachenmayer L, Heinemann U, Kluwe L, Finckh U, Höppner W, Stavrou D, Fillbrandt R, Westphal M. Molecular genetic alterations on chromosomes 11 and 22 in ependymomas. Int J Cancer. 2001;91(6):803–8.

301. McGuire CS, Sainani KL, Fisher PG. Incidence patterns for ependymoma: a surveillance, epidemiology, and end results study. J Neurosurg. 2009;110(4):725–9.

302. Kawabata Y, Takahashi JA, Arakawa Y, Hashimoto N. Long-term outcome in patients harboring intracranial ependymoma. J Neurosurg. 2005;103(1):31–7.

303. Sanford RA, Gajjar A. Ependymomas. Clin Neurosurg. 1997;44:559.

304. Mansur DB, Perry A, Rajaram V, Michalski JM, Park TS, Leonard JR, Luchtman-Jones L, Rich KM, Grigsby PW, Lockett MA, Wahab SH. Postoperative radiation therapy for grade II and III intracranial ependymoma. Int J Radiat Oncol Biol Phys. 2005;61(2):387–91.

305. Ammerman JM, Lonser RR, Dambrosia J, Butman JA, Oldfield EH. Long-term natural history of hemangioblastomas in patients with von Hippel-Lindau disease: implications for treatment. J Neurosurg. 2006;105(2):248–55.

306. Vortmeyer AO, Gnarra JR, Emmert-Buck MR, Katz D, Linehan WM, Oldfield EH, Zhuang Z. von Hippel-Lindau gene deletion detected in the stromal cell component of a cerebellar hemangioblastoma associated with von Hippel-Lindau disease. Hum Pathol. 1997;28(5):540–3.

307. Richard S, Campello C, Taillandier L, Parker F, Resche F. Haemangioblastoma of the central nervous system in von Hippel-Lindau disease. French VHL Study Group. J Intern Med. 1998;243(6):547–53.

308. Lonser RR, Glenn GM, Walther M, Chew EY, Libutti SK, Linehan WM, Oldfield EH. von Hippel-Lindau disease. Lancet. 2003;361(9374):2059–67.

309. Lonser RR, Butman JA, Kiringoda R, Song D, Oldfield EH. Pituitary stalk hemangioblastomas in von Hippel-Lindau disease. J Neurosurg. 2009;110(2):350–3.

310. Lonser RR, Butman JA, Huntoon K, Asthagiri AR, Wu T, Bakhtian KD, Chew EY, Zhuang Z, Linehan WM, Oldfield EH. Prospective natural history study of central nervous system hemangioblastomas in von Hippel-Lindau disease. J Neurosurg. 2014;120(5):1055–62.

311. Due-Tønnessen BJ, Helseth E, Scheie D, Skullerud K, Aamodt G, Lundar T. Long-term outcome after resection of benign cerebellar astrocytomas in children and young adults (0–19 years): report of 110 consecutive cases. Pediatr Neurosurg. 2002;37(2):71–80.

312. Sadighi Z, Slopis J. Pilocytic astrocytoma: a disease with evolving molecular heterogeneity. J Child Neurol. 2013;28(5):625–32.

313. El Beltagy MA, Atteya MM, El-Haddad A, Awad M, Taha H, Kamal M, Abou El Naga S. Surgical and clinical aspects of cerebellar pilomyxoid-spectrum astrocytomas in children. Childs Nerv Syst. 2014;30(6):1045–53.

314. Albright AL, Pollack IF, Andelson PD. Principles and practice of pediatric neurosurgery. Stuttgart: Thieme; 2015.

315. Poretti A, Meoded A, Cohen KJ, Grotzer MA, Boltshauser E, Huisman TA. Apparent diffusion coefficient of pediatric cerebellar tumors: a biomarker of tumor grade? Pediatr Blood Cancer. 2013;60(12):2036–41.

316. Steinbok P, Mangat JS, Kerr JM, Sargent M, Suryaningtyas W, Singhal A, Cochrane D. Neurological morbidity of surgical resection of pediatric cerebellar astrocytomas. Childs Nerv Syst. 2013;29(8):1269–75.

317. Wasilewski-Masker K, Liu Q, Yasui Y, Leisenring W, Meacham LR, Hammond S, Meadows AT, Robison LL, Mertens AC. Late recurrence in pediatric cancer: a report from the childhood cancer survivor study. J Natl Cancer Inst. 2009;101(24):1709–20.

318. Darrouzet V, Franco-Vidal V, Hilton M, Nguyen DQ, Lacher-Fougere S, Guerin J, Bebear JP. Surgery of cerebellopontine angle epidermoid cysts: role of the widened retrolabyrinthine approach combined with endoscopy. Otolaryngol Head Neck Surg. 2004;131(1):120–5.

319. Li F, Zhu S, Liu Y, Chen G, Chi L, Qu F. Hyperdense intracranial epidermoid cysts: a study of 15 cases. Acta Neurochir. 2007;149(1):31–9.

320. Takeshita M, Kubo O, Tajika Y, Izawa M, Kagawa M. Immunohistochemical detection of carbohydrate determinant 19–9 (CA 19–9) in intracranial epidermoid and dermoid cysts. Surg Neurol. 1993;40(4):284–8.

321. Akar Z, Tanriover N, Tuzgen S, Kafadar AM, Kuday C. Surgical treatment of intracranial epidermoid tumors. Neurol Med Chir. 2003;43(6):275–81.

322. Hoffman HJ, Becker L, Craven MA. A clinically and pathologically distinct group of benign brain stem gliomas. Neurosurgery. 1980;7(3):243–8.

323. Robertson PL, Allen JC, Abbott IR, Miller DC, Fidel J, Epstein FJ. Cervicomedullary tumors in children: a distinct subset of brainstem gliomas. Neurology. 1994;44(10):1798.

324. Klimo P, Goumnerova LC. Endoscopic third ventriculocisternostomy for brainstem tumors. J Neurosurg Pediatr. 2006;105(4):271–4.

325. Cornips EMJ, Overvliet GM, Weber JW, Postma AA, Hoeberigs CM, Baldewijns MMLL, et al. The clinical spectrum of Blake's pouch cyst: report of six illustrative cases. Childs Nerv Syst. 2010;26(8):1057–64.

326. Ramaswamy S, Rangasami R, Suresh S, Suresh I. Spontaneous resolution of Blake's pouch cyst. Radiol Case Rep. 2015;8(4):877.

327. Azab WA, Shohoud SA, Elmansoury TM, Salaheddin W, Nasim K, Parwez A. Blake's pouch cyst. Surg Neurol Int. 2014;5:112.

328. Kau T, Marterer R, Kottke R, Birnbacher R, Gellen J, Nagy E, Boltshauser E. Blake's pouch cysts and differential diagnoses in prenatal and postnatal MRI: a pictorial review. Clin Neuroradiol. 2020;30:435–45.

329. Liu Z, Han J, Fu F, Liu J, Li R, Yang X, et al. Outcome of isolated enlarged cisterna magna identified in utero: experience at a single medical center in mainland China. Prenat Diagn. 2017;37(6):575–82.

330. Bonde V, Muzumdar D, Goel A. Fourth ventricle arachnoid cyst. J Clin Neurosci. 2008;15(1):26–8.

331. Westermaier T, Vince GH, Meinhardt M, Monoranu C, Roosen K, Matthies C. Arachnoid cysts of the fourth ventricle—short illustrated review. Acta Neurochir. 2010;152(1):119–24.
332. Udayakumaran S, Biyani N, Rosenbaum DP, Ben-Sira L, Constantini S, Beni-Adani L. Posterior fossa craniotomy for trapped fourth ventricle in shunt-treated hydrocephalic children: long-term outcome. J Neurosurg Pediatr. 2011;7(1):52–63.
333. McClelland S 3rd, Ukwuoma OI, Lunos S, Okuyemi KS. The natural history of Dandy–Walker syndrome in the United States: a population-based analysis. J Neurosci Rural Pract. 2015;6(1):23–6.
334. Tadakamadla J, Kumar S, Mamatha GP. Dandy–Walker malformation: an incidental finding. Indian J Hum Genet. 2010;16(1):33–5.
335. Correa GG, Amaral LF, Vedolin LM. Neuroimaging of Dandy–Walker malformation: new concepts. Top Magn Reson Imaging. 2011;22(6):303–12.
336. Mohanty A, Biswas A, Satish S, Praharaj SS, Sastry KVR. Treatment options for Dandy–Walker malformation. J Neurosurg. 2006;105(5 Suppl):348–56.
337. Haddadi K, Zare A, Asadian L. Dandy–Walker syndrome: a review of new diagnosis and management in children. J Pediatr Rev. 2018;6(2):47–52.
338. Sasaki-Adams D, Elbabaa SK, Jewells V, Carter L, Campbell JW, Ritter AM. The Dandy–Walker variant: a case series of 24 pediatric patients and evaluation of associated anomalies, incidence of hydrocephalus, and developmental outcomes. J Neurosurg Pediatr. 2008;2(3):194–9.
339. Veretennikoff K, Coyne T, Biggs V, Robinson GA. Executive dysfunction after fourth-ventricle epidermoid cyst resection. Cogn Behav Neurol. 2018;31(4):207–13.
340. Tancredi A, Fiume D, Gazzeri G. Epidermoid cysts of the fourth ventricle: very long follow up in 9 cases and review of the literature. Acta Neurochir. 2003;145(10):901–5.
341. Chung LK, Beckett JS, Ong V, Lagman C, Nagasawa DT, Yang I, et al. Predictors of outcomes in fourth ventricular epidermoid cysts: a case report and a review of literature. World Neurosurg. 2017;105:689–96.
342. Simão D, Teixeira JC, Campos AR, Coiteiro D, Santos MM. Fourth ventricle neurocysticercosis: a case report. Surg Neurol Int. 2018;9:201.
343. Singh S, Gibikote SV, Shyamkumar NK. Isolated fourth ventricular cysticercus cyst: MR imaging in 4 cases with short literature review. Neurol India. 2003;51(3):394–6.
344. Haciyakupoglu E, Yilmaz DM, Kinali B, Akbas T, Haciyakupoglu S. Analysis of cavernous malformations: experience with 18 cases. Turk Neurosurg. 2019;29(3):340–8.
345. Polo G, Fischer C. Monitorage peropératoire des potentiels évoqués auditifs précoces dans la chirurgie de décompression microchirurgicale des nerfs crâniens de l'angle pontocérébelleux. Neurochirurgie. 2009;55(2):152–7.
346. Deletis V, Fernandez-Conejero I. Intraoperative monitoring and mapping of the functional integrity of the brainstem. J Clin Neurol. 2016;12(3):262–73.
347. Morota N, Deletis V. The importance of brainstem mapping in brainstem surgical anatomy before the fourth ventricle and implication for intraoperative neurophysiological mapping. Acta Neurochir (Wien). 2006;148(5):499–509.
348. Choque-Velasquez J, Hernesniemi J. One burr-hole craniotomy: suboccipital midline approach to the fourth ventricle in Helsinki neurosurgery. Surg Neurol Int. 2018;9:170.
349. Sakakibara F. Important points for patient positioning and head fixation in cerebral aneurysm surgery. No Shinkei geka Neurol Surg. 2021;49(1):24–30.

350. Klein O, Boussard N, Guerbouz R, Helleringer M, Joud A, Puget S. Surgical approach to the posterior fossa in children, including anesthetic considerations and complications: the prone and the sitting position. Technical note. Neurochirurgie. 2021;67(1):46–51.

351. Rozet I, Vavilala MS. Risks and benefits of patient positioning during neurosurgical care. Anesthesiol Clin. 2007;25(3):631–53.

352. Kwee MM, Ho YH, Rozen WM. The prone position during surgery and its complications: a systematic review and evidence-based guidelines. Int Surg. 2015;100(2):292–303.

353. Yilmazlar S, Aksoy K. Approach via the floor of the fourth ventricle for hydatid cyst of the pons. Pediatr Neurosurg. 1999;31(6):326–9.

354. Porter JM, Pidgeon C, Cunningham AJ. The sitting position in neurosurgery: a critical appraisal. Br J Anaesth. 1999;82(1):117–28.

355. Wang J, Wang X, Xu J, Wu Z, Dou Y. Microsurgical management of fourth ventricle lesions via the median suboccipital keyhole telovelar approach. J Craniofac Surg. 2022;15:10–97.

356. Hersh DS, Sanford KN, Moore K, Boop FA. Midline suboccipital craniotomy and direct stimulation for a dorsally exophytic brainstem tumor. Neurosurg Focus Video. 2019;1(2):V9.

357. Lieber S, Nunez M, Evangelista-Zamora R, Tatagiba M. Midline suboccipital subtonsillar approach with C1 laminectomy for resection of foramen magnum meningioma: 2-dimensional operative video. J Neurol Surg B Skull Base. 2019;80(Suppl 4):S365–S7.

358. Raheja A, Bowers CA, Couldwell WT. Midline telovelar and retrosigmoid suboccipital approaches for fenestration of multiple dilated Virchow–Robin spaces in the brainstem. Oper Neurosurg. 2017;13(5):645–6.

359. Varol E, Etli MU, Avcı F, Ramazanoğlu AF, Aydın SO, Yaltırık CK, Naderi S. Can posterior midline approach provide adequate exposure for all craniovertebral junction tumors? World Neurosurg. 2022;161:e482–7.

360. Panigrahi M, Samiruddin SM, Vooturi S. Crescent durotomy for midline posterior fossa lesions. Neurol India. 2019;67(1):155–8.

361. Tomasello F, Conti A, Cardali S, La Torre D, Angileri FF. Telovelar approach to fourth ventricle tumors: highlights and limitations. World Neurosurg. 2015;83(6):1141–7.

362. Ghali MGZ. Telovelar surgical approach. Neurosurg Rev. 2021;44(1):61–76.

363. Flannery AM, Mitchell L. Pediatric hydrocephalus: systematic literature review and evidence-based guidelines. Part 1: introduction and methodology. J Neurosurg Pediatr. 2014;14(Supplement_1):3–7.

364. Pillai SV. Techniques and nuances in ventriculoperitoneal shunt surgery. Neurol India. 2021;69(8):471.

365. Chung DY, Olson DM, John S, Mohamed W, Kumar MA, Thompson BB, Rordorf GA. Evidence-based management of external ventricular drains. Curr Neurol Neurosci Rep. 2019;19(12):94.

366. Maldonado IL, Valery CA, Boch AL. Shunt dependence: myths and facts. Acta Neurochir. 2010;152:1449–54.

367. Marino F, Simões AF, Simas Â, Pereira JG. Pneumocephalus following an accidental dural puncture, treated using hyperbaric oxygen therapy. A case report. J Crit Care Med. 2021;7(3):237–40.

368. Pulickal GG, Sitoh YY, Ng WH. Tension pneumocephalus. Singap Med J. 2014;55(3):e46.
369. Moskowitz SI, Liu J, Krishnaney AA. Postoperative complications associated with dural substitutes in suboccipital craniotomies. Neurosurgery. 2009;64(3):28–34.
370. Kaur H, Betances EM, Perera TB. Aseptic meningitis. Treasure Island (FL): StatPearls; 2023.
371. Shukla B, Aguilera EA, Salazar L, Wootton SH, Kaewpoowat Q, Hasbun R. Aseptic meningitis in adults and children: diagnostic and management challenges. J Clin Virol. 2017;94:110–4.
372. Simonin A, Levivier M, Bloch J, Messerer M. Cranial nerve palsies after shunting of an isolated fourth ventricle. Case Rep. 2015;2015:bcr2015209592.
373. de Laurentis C, Cristaldi PM, Rebora P, Valsecchi MG, Biassoni V, Schiavello E, Carrabba GG, Trezza A, DiMeco F, Ferroli P, Cinalli G. Posterior fossa syndrome in a population of children and young adults with medulloblastoma: a retrospective, multicenter Italian study on incidence and pathophysiology in a histologically homogeneous and consecutive series of 136 patients. J Neuro-Oncol. 2022;159(2):377–87.
374. Khan RB, Patay Z, Klimo P Jr, Huang J, Kumar R, Boop FA, Raches D, Conklin HM, Sharma R, Simmons A, Sadighi ZS. Clinical features, neurologic recovery, and risk factors of postoperative posterior fossa syndrome and delayed recovery: a prospective study. Neuro-Oncology. 2021;23(9):1586–96.
375. Wickenhauser ME, Khan RB, Raches D, Ashford JM, Robinson GW, Russell KM, et al. Characterizing posterior fossa syndrome: a survey of experts. Pediatr Neurol. 2020;104:19–22.
376. Delalande O, Rodriguez D, Chiron C, Fohlen M. Successful surgical relief of seizures associated with hamartoma of the floor of the fourth ventricle in children: report of two cases. Neurosurgery. 2001;49(3):726–31.
377. Jang SH, Yi JH, Kwon HG. Injury of the inferior cerebellar peduncle in patients with mild traumatic brain injury: a diffusion tensor tractography study. Brain Inj. 2016;30(10):1271–5.
378. de Groat WC. Anatomy of the central neural pathways controlling the lower urinary tract. Eur Urol. 1998;34(Suppl. 1):2–5.
379. Bartlett F, Kortmann R, Saran F. Medulloblastoma. Clin Oncol. 2013;25(1):36–45.
380. Jiang T, Zhang Y, Wang J, Du J, Ma Z, Li C, Liu R, Zhang Y. Impact of tumor location and fourth ventricle infiltration in medulloblastoma. Acta Neurochir. 2016;158:1187–95.
381. Srinivasan VM, Ghali MG, North RY, Boghani Z, Hansen D, Lam S. Modern management of medulloblastoma: molecular classification, outcomes, and the role of surgery. Surg Neurol Int. 2016;7(Suppl 44):S1135–S41.
382. Ghali MG. Microsurgical techniques for achieving gross total resection of ependymomas of the fourth ventricle. Acta Chir Belg. 2020;120(3):149–66.
383. Winkler EA, Birk H, Safaee M, Yue JK, Burke JF, Viner JA, Pekmezci M, Perry A, Aghi MK, Berger MS, McDermott MW. Surgical resection of fourth ventricular ependymomas: case series and technical nuances. J Neuro-Oncol. 2016;130:341–9.
384. Sabbagh AJ, Alaqeel AM. Focal brainstem gliomas: advances in intra-operative management. Neurosci J. 2015;20(2):98–106.
385. Wang F, Su X, Zhang W, Song J. How I do it? The surgical resection of the fourth ventricle choroid plexus papilloma. Acta Neurochir. 2022;165(7):1767–71.

386. Hosmann A, Hinker F, Dorfer C, Slavc I, Haberler C, Dieckmann K, et al. Management of choroid plexus tumors-an institutional experience. Acta Neurochir. 2019;161(4):745–54.

387. Chen CM, Chen YC, Chu PY, Kuo LT. A cystic lesion in the fourth ventricle. J Clin Neurosci. 2011;18(9):1282.

388. Nishimoto A, Kawakami Y. Surgical removal of hemangioblastoma in the fourth ventricle. Surg Neurol. 1980;13(6):423–7.

389. Franko LR, Pandian B, Gupta A, Savastano LE, Chen KS, Riddell J IV, Orringer DA. Posterior fossa craniotomy for adherent fourth ventricle neurocysticercosis. Oper Neurosurg. 2019;16(5):E154–8.

390. Krishnamurthy S, Li J. New concepts in the pathogenesis of hydrocephalus. Transl Pediatr. 2014;3(3):185.

391. Koga H, Mori K, Kawano T, Tsutsumi K, Jinnouchi T. Parinaud's syndrome in hydrocephalus due to a basilar artery aneurysm. Surg Neurol. 1983;19(6):548–53.

392. Williams MA, Malm J. Diagnosis and treatment of idiopathic normal pressure hydrocephalus. Continuum (Minneap Minn). 2016;22(2 Dementia):579–99.

393. Zahl SM, Egge A, Helseth E, Wester K. Benign external hydrocephalus: a review, with emphasis on management. Neurosurg Rev. 2011;34:417–32.

394. Bradley WG Jr. Diagnostic tools in hydrocephalus. Neurosurg Clin N Am. 2001;12(4):661–84.

395. Damasceno BP. Normal pressure hydrocephalus: diagnostic and predictive evaluation. Dementia Neuropsychol. 2009;3:8–15.

396. Ruggiero C, Cinalli G, Spennato P, Aliberti F, Cianciulli E, Trischitta V, Maggi G. Endoscopic third ventriculostomy in the treatment of hydrocephalus in posterior fossa tumors in children. Childs Nerv Syst. 2004;20:828–33.

397. Bagga R, Jain V, Gupta KR, Gopalan S, Malhotra S, Das CP. Choice of therapy and mode of delivery in idiopathic intracranial hypertension during pregnancy. Medscape Gen Med. 2005;7(4):42.

398. de Melo Mussi AC, de Oliveira EP. Ventricular anatomy. In: Comprehensive overview of modern surgical approaches to intrinsic brain tumors. Academic Press; 2019. p. 107–18.

399. Miller J, Hdeib A, Cohen A. Management of tumors of the fourth ventricle. In: Schmidek and sweet operative neurosurgical techniques: indications, methods, and results. 6th ed. Amsterdam: Elsevier Inc.; 2012. p. 367–97.

400. Akakin A, Peris-Celda M, Kilic T, Seker A, Gutierrez-Martin A, Rhoton A Jr. The dentate nucleus and its projection system in the human cerebellum: the dentate nucleus microsurgical anatomical study. Neurosurgery. 2014;74(4):401–25.

401. Rhoton AL Jr. Cerebellum and fourth ventricle. Neurosurgery. 2000;47(suppl_3):S7–27.

402. Paff M, Alexandru-Abrams D, Muhonen M, Loudon W. Ventriculoperitoneal shunt complications: a review. Interdiscip Neurosurg. 2018;13:66–70.

403. Pan P. Outcome analysis of ventriculoperitoneal shunt surgery in pediatric hydrocephalus. J Pediatr Neurosci. 2018;13(2):176.

404. Hultegård L, Michaëlsson I, Jakola A, Farahmand D. The risk of ventricular catheter misplacement and intracerebral hemorrhage in shunt surgery for hydrocephalus. Interdiscip Neurosurg. 2019;17:23–7.

405. Lee TT, Uribe J, Ragheb J, Morrison G, Jagid JR. Unique clinical presentation of pediatric shunt malfunction. Pediatr Neurosurg. 1999;30(3):122–6.

406. Yamada S, Kitagawa R, Teramoto A. Relationship of the location of the ventricular catheter tip and function of the ventriculoperitoneal shunt. J Clin Neurosci. 2013;20(1):99–101.

407. Wall M. Idiopathic intracranial hypertension. Neurol Clin. 2010;28(3):593–617.

408. Bari M, Khan F, Rehman A, Shamim M. Factors affecting ventriculoperitoneal shunt survival in adult patients. Surg Neurol Int. 2015;6(1):25.

409. Won S, Dubinski D, Behmanesh B, Bernstock J, Seifert V, Konczalla J, et al. Management of hydrocephalus after resection of posterior fossa lesions in pediatric and adult patients—predictors for development of hydrocephalus. Neurosurg Rev. 2019;43(4):1143–50.

410. Fowler JB, De Jesus O, Mesfin FB. Ventriculoperitoneal shunt. Treasure Island (FL): StatPearls; 2023.

411. Rajshekhar V, Moorthy R. Endoscopic third ventriculostomy for hydrocephalus: a review of indications, outcomes, and complications. Neurol India. 2011;59(6):848.

412. Aoki N. Lumboperitoneal shunt: clinical applications, complications, and comparison with ventriculoperitoneal shunt. Neurosurgery. 1990;26(6):998–1004.

413. Yadav Y, Parihar V, Sinha M. Lumbar peritoneal shunt. Neurol India. 2010;58(2):179.

414. Singh R, Prasad RS, Singh RC, Trivedi A, Bhaikhel KS, Sahu A. Evaluation of pediatric hydrocephalus: clinical, surgical, and outcome perspective in a tertiary center. Asian J Neurosurg. 2021;16(4):706–13.

415. Beucler N, Hibbert D, Vioujard C, Joubert C, Dagain A. When ventriculoatrial shunt gets lost. Am J Interv Radiol. 2020;4:1.

416. Nee L, Harun R, Sellamuthu P, Idris Z. Comparison between ventriculosubgaleal shunt and extraventricular drainage to treat acute hydrocephalus in adults. Asian J Neurosurg. 2017;12(4):659.

417. Kormanik K, Praca J, Garton H, Sarkar S. Repeated tapping of ventricular reservoir in preterm infants with post-hemorrhagic ventricular dilatation does not increase the risk of reservoir infection. J Perinatol. 2009;30(3):218–21.

418. Kline C, Packer R, Hwang E, Raleigh D, Braunstein S, Raffel C, et al. Case-based review: pediatric medulloblastoma. Neurooncol Pract. 2017;4(3):138–50.

419. Hasslacher-Arellano J, Arellano-Aguilar G, Funes-Rodríguez J, López-Forcén S, Torres-Zapiain F, Domínguez-Carrillo L. Ventriculo-gallbladder shunt: an alternative for the treatment of hydrocephalus. Cirugía y Cirujanos (English Edition). 2016;84(3):225–9.

420. Das JM, Biagioni MC. Normal pressure hydrocephalus. Treasure Island (FL): StatPearls; 2023.

421. Foreman P, Hendrix P, Griessenauer C, Schmalz P, Harrigan M. External ventricular drain placement in the intensive care unit versus operating room: evaluation of complications and accuracy. Clin Neurol Neurosurg. 2015;128:94–100.

422. Hochstetler A, Raskin J, Blazer-Yost BL. Hydrocephalus: historical analysis and considerations for treatment. Eur J Med Res. 2022;27(1):168.

423. Al-Ameri L. Brain endoscopy, a big neurosurgical revolution. Al-Kindy Coll Med J. 2017;13(2):1–5.

424. Yadav Y, Parihar V, Pande S, Namdev H, Agarwal M. Endoscopic third ventriculostomy. J Neurosci Rural Pract. 2012;3(02):163–73.

425. Zheng J, Chen G, Xiao Q, Huang Y, Guo Y. Endoscopy in the treatment of slit ventricle syndrome. Exp Ther Med. 2017;14(4):3381–6.

426. Sadler TW. Langman's medical embryology. 12th ed. Philadelphia: Lippincott Williams and Wilkins; 2012.

427. Pal, B. R., Preston, P. R., Morgan, M. E., Rushton, D. I., and, Durbin, G. M. (2001). Frontal horn thin walled cysts in preterm neonates are benign. Arch Dis Child Fetal Neonatal Ed, 85(3), F187–F193. https://doi.org/10.1136/fn.85.3.f187

428. Marks G. Hydrocephalus and cerebrospinal fluid (CSF). In: Greenberg's handbook of neurosurgery. 9th ed. New York: Thieme; 2023. p. 415–6.

429. Rodriguez Criado G, Perez Aytes A, Martinez F, Vos YJ, Verlind E, Gonzalez-Meneses Lopez A, et al. X-linked hydrocephalus: another two families with an L1 mutation. Genet Couns. 2003;14(1):57–65.

430. Stumpel C, Vos YJ. L1 syndrome. In: Adam MP, Ardinger HH, Pagon RA, et al., editors. GeneReviews®. Seattle: University of Washington; 2004.

431. Lopez JA, Reich D. Choroid plexus cysts. J Am Board Fam Med. 2006;19(4):422–5. https://doi.org/10.3122/jabfm.19.4.422.

432. Longatti P, Fiorindi A, Perin A, Martinuzzi A. Endoscopic anatomy of the cerebral aqueduct. Neurosurgery. 2007;61(3 Suppl):1–6.

433. Lewis HK. Observations on the pathology of hydrocephalus. Ulster Med J. 1949;18(1):107–8.

434. Kaur LP, Munyiri NJ, Dismus WV. Clinical analysis of aqueductal stenosis in patients with hydrocephalus in a Kenyan setting. Pan Afr Med J. 2017;26:106.

435. Zamora EA, Dandy AT. Walker malformation. Treasure Island (FL): StatPearls; 2023.

436. Zamora EA, Ahmad T. Dandy–Walker malformation. In: StatPearls. Treasure Island (FL): StatPearls Publishing; 2020.

437. Blank MC, Grinberg I, Aryee E, et al. Multiple developmental programs are altered by loss of Zic1 and Zic4 to cause Dandy–Walker malformation cerebellar pathogenesis. Development. 2011;138(6):1207–16.

438. Oria MS, Rasib AR, Pirzad AF, Wali Ibrahim Khel F, Ibrahim Khel MI, Wardak FR. A rare case of Dandy–Walker syndrome. Int Med Case Rep J. 2022;15:55–9.

439. Sun Y, Wang T, Zhang N, Zhang P, Li Y. Clinical features and genetic analysis of Dandy–Walker syndrome. BMC Pregnancy Childbirth. 2023;23(1):40.

440. Abebe MS, Seyoum G, Emamu B, Teshome D. Congenital hydrocephalus and associated risk factors: an institution-based case-control study, Dessie Town, North East Ethiopia. Pediatr Health Med Ther. 2022;13:175–82.

441. Cinalli G, Spennato P, Nastro A, Aliberti F, Trischitta V, Ruggiero C, et al. Hydrocephalus in aqueductal stenosis. Childs Nerv Syst. 2011;27(10):1621–42.

442. Heaphy-Henault KJ, Guimaraes CV, Mehollin-Ray AR, Cassady CI, Zhang W, Desai NK, et al. Congenital aqueductal stenosis: findings at fetal MRI that accurately predict a postnatal diagnosis. AJNR Am J Neuroradiol. 2018;39(5):942–8.

443. Sutton LN, Bruce DA, Schut L. Hydranencephaly versus maximal hydrocephalus: an important clinical distinction. Neurosurgery. 1980;6(1):34–8.

444. DeMyer W. Median facial malformations and their implications for brain malformations. Birth Defects Orig Artic Ser. 1975;11(7):155–81.

445. Kular S, Cascella M. Chiari I malformation. Treasure Island (FL): StatPearls Publishing; 2023.

446. Langridge B, Phillips E, Choi D. Chiari malformation type 1: a systematic review of natural history and conservative management. World Neurosurg. 2017;104:213–9.

447. Søvik O, van der Hagen CB, Løken HC. X-linked aqueductal stenosis. Clin Genet. 1977;11(5):416–20.

448. Hochberg E, Niles E. An incidental finding of Dandy–Walker malformation. JAAPAs. 2021;34(1):22–4.

449. Hidalgo JA, Tork CA, Varacallo M. Arnold Chiari malformation. Treasure Island (FL): StatPearls; 2023.

450. Schneider B, Birthi P, Salles S. Arnold-Chiari 1 malformation type 1 with syringohydromyelia presenting as acute tetraparesis: a case report. J Spinal Cord Med. 2013;36(2):161–5.

451. Guo F, Turgut M. Precise management of Chiari malformation with type I. Front Surg. 2022;9:850879.

452. Jadhav SS, Dhok A, Mitra K, Khan S, Khandaitkar S. Dandy–Walker malformation with hydrocephalus: diagnosis and its treatment. Cureus. 2022;14(5):e25287.

453. Cousins O, Hodges A, Schubert J, Veronese M, Turkheimer F, Miyan J, Engelhardt B, Roncaroli F. The blood–CSF–brain route of neurological disease: the indirect pathway into the brain. Neuropathol Appl Neurobiol. 2022;48(4):e12789.

454. Winn HR. Youmans and Winn neurological surgery e-book. Elsevier Health Sciences; 2022.

455. Tunkel AR, Hasbun R, et al. Infectious Diseases Society of America clinical practice guidelines for healthcare-associated ventriculitis and meningitis. Clin Infect Dis. 2017;64(6):e34–65.

456. Fukui MB, Williams RL, Mudigonda S. CT and MR imaging features of pyogenic ventriculitis. Am J Neuroradiol. 2001;22(8):1510–6.

457. Fujikawa A, Tsuchiya K, Honya K, Nitatori T. Comparison of MRI sequences to detect ventriculitis. Am J Roentgenol. 2006;187(4):1048–53.

458. Martin RM, Zimmermann LL, Huynh M, Polage CR. Diagnostic approach to health care-and device-associated central nervous system infections. J Clin Microbiol. 2018;56(11):e00861–18.

459. Korinek AM, Baugnon T, Golmard JL, van Effenterre R, Coriat P, Puybasset L. Risk factors for adult nosocomial meningitis after craniotomy: role of antibiotic prophylaxis. Neurosurgery. 2006;59(1):126–33.

460. Chiang HY, Kamath AS, Pottinger JM, Greenlee JD, Howard MA, Cavanaugh JE, Herwaldt LA. Risk factors and outcomes associated with surgical site infections after craniotomy or craniectomy. J Neurosurg. 2014;120(2):509–21.

461. Gadgil N, Chamoun RB, Gopinath SP. Intraventricular brain abscess. J Clin Neurosci. 2012;19(9):1314–6.

462. Lee TH, Chang WN, Su TM, Chang HW, Lui CC, Ho JT, Wang HC, Lu CH. Clinical features and predictive factors of intraventricular rupture in patients who have bacterial brain abscesses. J Neurol Neurosurg Psychiatry. 2007;78(3):303–9.

463. Bhatia PL, Lilani S, Shirpurkar R, Chande C, Joshi S, Chowdhary A. Coagulase-negative staphylococci: emerging pathogen in central nervous system shunt infection. Indian J Med Microbiol. 2017;35(1):120–3.

464. Raffa G, Marseglia L, Gitto E, Germanò A. Antibiotic-impregnated catheters reduce ventriculoperitoneal shunt infection rate in high-risk newborns and infants. Childs Nerv Syst. 2015;31:1129–38.

465. Fan-Havard P, Nahata MC. Treatment and prevention of infections of cerebrospinal fluid shunts. Clin Pharm. 1987;6(11):866–80.

466. Forte D, Peraio S, Huttunen TJ, James G, Thompson D, Aquilina K. Ventriculoatrial and ventriculopleural shunts as second-line surgical treatment have equivalent revision,

infection, and survival rates in paediatric hydrocephalus. Childs Nerv Syst. 2021;37: 481–9.

467. Lozier AP, Sciacca RR, Romagnoli MF, Connolly ES Jr. Ventriculostomy-related infections: a critical review of the literature. Neurosurgery. 2002;51(1):170–82.

468. Nilsson A, Uvelius E, Cederberg D, Kronvall E. Silver-coated ventriculostomy catheters do not reduce rates of clinically diagnosed ventriculitis. World Neurosurg. 2018;117:e411–6.

469. Beer R, Lackner P, Pfausler B, Schmutzhard E. Nosocomial ventriculitis and meningitis in neurocritical care patients. J Neurol. 2008;255(11):1617–24.

470. Karvouniaris M, Brotis A, Tsiakos K, Palli E, Koulenti D. Current perspectives on the diagnosis and management of healthcare-associated ventriculitis and meningitis. Infect Drug Resist. 2022;15:697–721.

471. Bischoff P, Schroder C, Gastmeier P, Geffers C. Surveillance of external ventricular drainage-associated meningitis and ventriculitis in German intensive care units. Infect Control Hosp Epidemiol. 2020;41(4):452–7.

472. Wada T, Kuroda K, Yoshida Y, Moriguchi T, Nishikawa Y, Ogawa A, Taniguchi S, Saitoh K. A case of posttraumatic severe ventriculitis treated by intraventricular lavage. No Shinkei geka Neurol Surg. 2000;28(8):737–43.

473. Tunkel AR, Hartman BJ, Kaplan SL, et al. Practice guidelines for the management of bacterial meningitis. Clin Infect Dis. 2004;39:1267–84.

474. Shenoy SS, Lui F. Neuroanatomy, ventricular system. Treasure Island (FL): StatPearls; 2023.

475. Vinas FC, Lopez F, Dujovny M. Microsurgical anatomy of the posterior choroidal arteries. Neurol Res. 1995;17(5):334–44.

476. Xu Y, Mohyeldin A, Doniz-Gonzalez A, Vigo V, Pastor-Escartin F, Meng L, Cohen-Gadol AA, Fernandez-Miranda JC. Microsurgical anatomy of the lateral posterior choroidal artery: implications for intraventricular surgery involving the choroid plexus. J Neurosurg. 2021;135(5):1534–49.

477. Javed K, Das JM. Neuroanatomy, anterior choroidal arteries. Treasure Island (FL): StatPearls; 2023.

478. Morandi X, Brassier G, Darnault P, Mercier P, Scarabin JM, Duval JM. Anatomie microchirurgicale de l'artère choroidienne antérieure. Surg Radiol Anat. 1996;18:275–80.

479. Javed K, Reddy V, Das JM. Neuroanatomy, posterior cerebral arteries. Treasure Island (FL): StatPearls; 2023.

480. Grigoryan YA, Sitnikov AR, Timoshenkov AV, Grigoryan GY. Anevrizma medial'noĭ zadneĭ vorsinchatoĭ arterii: opisanie klinicheskogo nabliudeniia i Obzor literatury [an aneurysm of the medial posterior choroidal artery: a case report and a literature review]. Zh Vopr Neirokhir Im N N Burdenko. 2017;81(4):101–7.

481. Kılıç T, Akakın A. Anatomy of cerebral veins and sinuses. Front Neurol Neurosci. 2008;23:4–15.

482. Uddin MA, Haq TU, Rafique MZ. Cerebral venous system anatomy. J Pak Med Assoc. 2006;56(11):516.

483. Azab WA, Salaheddin W, Alsheikh TM, Nasim K, Nasr MM. Colloid cysts posterior and anterior to the foramen of Monro: anatomical features and implications for endoscopic excision. Surg Neurol Int. 2014;5:124.

484. Peltier J, Baroncini M, Zunon-Kipré Y, Haidara A, Havet E, Foulon P, Page C, Lejeune JP, Le Gars D. Vascularisation des ventricules latéraux. Neurochirurgie. 2011;57(4–6):156–60.

485. Scelsi CL, Rahim TA, Morris JA, Kramer GJ, Gilbert BC, Forseen SE. The lateral ventricles: a detailed review of anatomy, development, and anatomic variations. AJNR Am J Neuroradiol. 2020;41(4):566–72.

486. Chen Z, Qiao H, Guo Y, Li J, Miao H, Wen C, Wen X, Zhang X, Yang X, Chen C. Visualization of anatomic variation of the anterior septal vein on susceptibility-weighted imaging. PLoS One. 2016;11(10):e0164221.

487. Türe U, Ekinci G, Necmettin Pamir M, Erzen C. Venous variations in the region of the third ventricle: the role of MR venography. Neuroradiology. 2003;45(12):900.

488. Leonardo J, Grand W. Enlarged thalamostriate vein causing unilateral Monro foramen obstruction: case report. J Neurosurg Pediatr. 2009;3(6):507–10.

489. Tubbs RS, Loukas M, Louis RG, Shoja MM, Askew CS, Phantana-Angkool A, Salter EG, Oakes WJ. Surgical anatomy and landmarks for the basal vein of Rosenthal. J Neurosurg. 2007;106(5):900–2.

490. Brzegowy K, Zarzecki MP, Musial A, Aziz HM, Kasprzycki T, Tubbs RS, et al. The internal cerebral vein: new classification of branching patterns based on CTA. AJNR Am J Neuroradiol. 2019;40(10):1719–24.

491. Miao HL, Zhang DY, Wang T, Jiao XT, Jiao LQ. Clinical importance of the posterior inferior cerebellar artery: a review of the literature. Int J Med Sci. 2020;17(18):3005.

492. Ballabh P. Intraventricular hemorrhage in premature infants: mechanism of disease. Pediatr Res. 2010;67(1):1–8.

493. Angelopoulos M, Gupta SR. Primary intraventricular hemorrhage in adults: clinical features, risk factors, and outcome. Surg Neurol. 1995;44(5):433–6.

494. Mohr G, Ferguson G, Khan M, Malloy D, Watts R, Benoit B, Weir B. Intraventricular hemorrhage from ruptured aneurysm: retrospective analysis of 91 cases. J Neurosurg. 1983;58(4):482–7.

495. Stretz C, Sheikh Z, Maciel CB, Hirsch LJ, Gilmore EJ. Seizures, periodic and rhythmic patterns in primary intraventricular hemorrhage. Ann Clin Transl Neurol. 2018;5(9):1104–11.

496. Flint AC, Roebken A, Singh V. Primary intraventricular hemorrhage: yield of diagnostic angiography and clinical outcome. Neurocrit Care. 2008;8:330–6.

497. Anderson CS, Heeley E, Huang Y, Wang J, Stapf C, Delcourt C, Lindley R, Robinson T, Lavados P, Neal B, Hata J. Rapid blood-pressure lowering in patients with acute intracerebral hemorrhage. N Engl J Med. 2013;368:2355–65.

498. Morgenstern LB, Hemphill JC III, Anderson C, Becker K, Broderick JP, Connolly ES Jr, Greenberg SM, Huang JN, Macdonald RL, Messé SR, Mitchell PH. Guidelines for the management of spontaneous intracerebral hemorrhage: a guideline for healthcare professionals from the American Heart Association/American Stroke Association. Stroke. 2010;41(9):2108–29.

499. Darby DG, Donnan GA, Saling MA, Walsh KW, Bladin PF. Primary intraventricular hemorrhage: clinical and neuropsychological findings in a prospective stroke series. Neurology. 1988;38(1):68–75.

500. Volpe JJ. Neonatal intraventricular hemorrhage. N Engl J Med. 1981;304(15):886–91.

501. Szpecht D, Frydryszak D, Miszczyk N, Szymankiewicz M, Gadzinowski J. The incidence of severe intraventricular hemorrhage based on retrospective analysis of 35,939

full-term newborns—report of two cases and review of literature. Childs Nerv Syst. 2016;32(12):2447–51.

502. Ballabh P, Braun A, Nedergaard M. Anatomic analysis of blood vessels in germinal matrix, cerebral cortex, and white matter in developing infants. Pediatr Res. 2004;56(1):117–24.

503. Amato M, Howald H, Von Muralt G. Fetal sex and distribution of peri-intraventricular hemorrhage in preterm infants. Eur Neurol. 1987;27(1):20–3.

504. Hambleton G, Wigglesworth JS. Origin of intraventricular haemorrhage in the preterm infant. Arch Dis Child. 1976;51(9):651–9.

505. Wirtschafter DD, Danielsen BH, Main EK, Korst LM, Gregory KD, Wertz A, Stevenson DK, Gould JB, Collaborative CP. Promoting antenatal steroid use for fetal maturation: results from the California Perinatal Quality Care Collaborative. J Pediatr. 2006;148(5):606–12.

506. Papile LA, Burstein J, Burstein R, Koffler H. Incidence and evolution of subependymal and intraventricular hemorrhage: a study of infants with birth weights less than 1500 g. J Pediatr. 1978;92(4):529–34.

507. Castillo M. Neuroradiology companion: methods, guidelines, and imaging fundamentals. Philadelphia: Lippincott Williams and Wilkins; 2006.

508. Hill A, Rozdilsky B. Congenital hydrocephalus secondary to intra-uterine germinal matrix/intraventricular haemorrhage. Dev Med Child Neurol. 1984;26(4):524–7.

509. International PHVD Drug Trial Group. International randomised controlled trial of acetazolamide and furosemide in posthaemorrhagic ventricular dilatation in infancy. Lancet. 1998;352(9126):433–40.

510. Kreusser KL, Tarby TJ, Kovnar E, Taylor DA, Hill A, Volpe JJ. Serial lumbar punctures for at least temporary amelioration of neonatal posthemorrhagic hydrocephalus. Pediatrics. 1985;75(4):719–24.

511. Hudgins RJ, Boydston WR, Gilreath CL. Treatment of posthemorrhagic hydrocephalus in the preterm infant with a ventricular access device. Pediatr Neurosurg. 1998;29(6):309–13.

512. Rahman S, Teo C, Morris W, Lao D, Boop FA. Ventriculosubgaleal shunt: a treatment option for progressive posthemorrhagic hydrocephalus. Childs Nerv Syst. 1995;11(11):650–4.

513. Nehls DG, Flom RA, Carter LP, Spetzler RF. Multiple intracranial aneurysms: determining the site of rupture. J Neurosurg. 1985;63(3):342–8.

514. Yeh HS, Tomsick TA, Tew JM. Intraventricular hemorrhage due to aneurysms of the distal posterior inferior cerebellar artery: report of three cases. J Neurosurg. 1985;62(5):772–5.

515. Yuan Z, Woha Z, Weiming X. Intraventricular aneurysms: case reports and review of the literature. Clin Neurol Neurosurg. 2013;115(1):57–64.

516. Kivelev J, Niemelä M, Kivisaari R, Hernesniemi J. Intraventricular cerebral cavernomas: a series of 12 patients and review of the literature. J Neurosurg. 2010;112(1): 140–9.

517. Graupman P, Defillo A, Nussbaum L, Nussbaum ES. Limited vertical dural opening for lesions of the vermis, fourth ventricle, and distal PICA segments. Surg Neurol Int. 2012;3:141.

518. Gupta N, Grover H, Bansal I, Hooda K, Sapire JM, Anand R, Kumar Y. Neonatal cranial sonography: ultrasound findings in neonatal meningitis—a pictorial review. Quant Imaging Med Surg. 2017;7(1):123.

MIX
Papier aus verantwortungsvollen Quellen
Paper from responsible sources
FSC® C105338
www.fsc.org

If you have any concerns about our products,
you can contact us on
**ProductSafety@springernature.com**

In case Publisher is established outside the EU,
the EU authorized representative is:
**Springer Nature Customer Service Center GmbH
Europaplatz 3, 69115 Heidelberg, Germany**

Printed by Libri Plureos GmbH
in Hamburg, Germany